WHAT CAN NANOTECHNOLOGY LEARN FROM BIOTECHNOLOGY?

Food Science and Technology
International Series

A complete list of books in this series appears at the end of this volume.

What Can Nanotechnology Learn from Biotechnology?

Social and Ethical Lessons for Nanoscience from the Debate over Agrifood Biotechnology and GMOs

Edited by

Kenneth David, Ph.D.
Department of Anthropology,
Michigan State University

Paul B. Thompson, Ph.D.
Departments of Philosophy,
Agricultural Economics, and Community, Agriculture,
Recreation and Resource Studies,
Michigan State University

AMSTERDAM • BOSTON • HEIDELBERG • LONDON • NEW YORK • OXFORD
PARIS • SAN DIEGO • SAN FRANCISCO • SINGAPORE • SYDNEY • TOKYO
Academic Press is an imprint of Elsevier

Academic Press is an imprint of Elsevier
30 Corporate Drive, Suite 400, Burlington, MA 01803, USA
84 Theobald's Road, London WC1X 8RR, UK
525 B Street, Suite 1900, San Diego, California 92101-4495, USA

First edition 2008

Library of Congress Cataloging in Publication Data
A catalog record for this book is available from the Library of Congress

British Library Cataloguing in Publication Data
A catalogue record for this book is available from the British Library

ISBN: 978-012-373990-2

For information on all Academic Press publications
visit our web site at books.elsevier.com

Typeset by Charon Tec Ltd (A Macmillan Company), Chennai, India
www.charontec.com

Printed and bound in the United States of America

08 09 10 11 12 10   9 8 7 6 5 4 3 2 1

Contents

Acknowledgments

We would like to thank the National Science Foundation (NSF).[1] Both the 2005 "What Can Nano Learn from Bio?" International Conference and Workshop and the 2006 "Standards for Nanotechnology" workshop were supported by a National Science Foundation/Nanoscale Interdisciplinary Research Teams Grant: Building Capacity for Social and Ethical Research and Education in Agrifood Nanotechnology (SES-0403847). In particular, we give thanks to Rachelle Hollander (retired), Mihail Roco, Rita Teutonico, and Priscilla Regan of NSF. We further thank the W.K. Kellogg Foundation for support extended through the W.K. Kellogg Endowed Professorship in Agricultural, Food and Community Ethics.

The 2005 international conference was sparked by opening remarks from Lou Anna K. Simon, President of Michigan State University.

Event planning and report preparation are truly collaborative efforts and this workshop was no exception. Special thanks are due the following people for their roles in helping to plan, organize, and host the 2005 conference and the 2006 workshop: Julie Eckinger, Sibbir Noman, Scott Menhart, Jim Sumbler, Linda Estill, Mary Keyes, Nicole Schoendorf, and Jill Crandell for administrative, logistical and secretarial support; Brian Cools and Linda Currier for graphic design.

Event delivery at the 2005 Conference—convened at Kellogg Hotel & Conference Center: Tammi J. Cady, Rhonda Bucholtz, and

---

1. The views expressed here are those of the conference and workshop participants and reviewers, and do not necessarily reflect those of the National Science Foundation, Michigan State University, or the participants' employers.

Bill Burke. Event delivery at the 2006 Workshop—convened at Cowles House: Peter Lechloer and Theresa Pharms.

Thanks to graduate research assistants Sho (Lisa) Ngai, Meghan Sullivan, Norbismi Nordin, Brian Depew, Zahra Meghani, and William Hannah and undergraduate research assistants Lawrence Judd, Erin Pullen, and Keiko Tanaka for their help in taking notes and recordings made at the conference and workshops. This material made a contribution to both the Introduction and the Standards workshop report. John Stone had primary responsibility for assembling the report. Thanks to Marc Erbisch and Erin Pullen for many hours of help in preparing the chapter manuscripts for publication. Further thanks to Nancy Maragioglio, Editor and Claire Hutchins, Project Manager from Academic/Elsevier who worked closely and effectively with us in producing this volume.

Thanks to Agrifood Nanotechnology Project research project members Les Bourquin, Lawrence Busch, Kenneth David, Brady Deaton, Tom Dietz, John Lloyd, Susan Selke, John Stone, Deepa Thiagarajan, and Paul Thompson; the Department of Sociology, the Department of Anthropology, the Department of Community, Agriculture, Recreation and Resource Studies, the Department of Packaging, the College of Social Science and the Michigan Agricultural Experiment Station—all at Michigan State University; and, of course, thanks to the workshop attendees without whose enthusiastic participation the workshop would not have been possible.

Preface

The rapid growth of nanoscience and nanotechnology is a global and widely acknowledged phenomenon. In Europe and the United States in particular, the rapid increase in both public and private investment in nono-scale science and technology has been accompanied by statements recognizing the need to steer the process in a democratic fashion and to secure broad public acceptance. The international controversy over genetically engineered crops and livestock is often mentioned in this connection. Commentators from industry, government and public interest organizations alike pledge to "learn the lessons," from the successes and failures of scientists, regulators and companies who developed the technology that came to be popularly known as "genetically modified organisms," or GMOs.

But what were those lessons? This volume is the result of a systematically planned research activity designed to answer that question. To that end, the editors and several colleagues at Michigan State University undertook a three year process to survey literature on the GMO controversy, contact a number of authors who had made distinguished contributions to that literature, and to bring them together in a workshop setting with others who were undertaking both technical applications in nano-scale science and engineering as well as schlorship on the processes of governance and public acceptance of nanotechnology. This volume is the end product of that research, consisting of reflective and critical essays written by just a few of the participants in this iterative interdisciplinary research project. We owe an enormous debt to all of those who participated in our workshop, as well as to all the members of Michigan State Agrifood Nanotechnology Research Team responsible for planning and conducting the research. Research assistants for the project were especially important in actually making the nuts and bolts of the conference and workshop work. These names are listed in the acknowledgments and in appendices to the volume.

We would like, however, to make special note of the career contribution that Dr. Rachelle Hollander has made to research on the social and ethical issues in science and engineering. Her important research contributions speak for themselves. What may be less evident to outsiders is the continuing role that she played at the National Science Foundation in finding an institutional home for this work, not to mention dollars to support it. Her last assignment at NSF before entering what we hope will be a well earned but still productive retirement was to help lay the foundations for the program in Social and Ethical Issues in Nanotechnology component of the National Nanotechnology Initiative. Without that work, this volume would truly have been impossible. It is to Rachelle that this book is dedicated.

About the Authors

David J. Bjornstad, Society-Technology Interactions Group, Environmental Sciences Division, Oak Ridge National Laboratory, Oak Ridge, Tennessee. His research centers on the economic policy analysis on topics dealing with science policy and energy environment and natural resources policy, applied microeconomic theory, natural resource valuation, and experimental economics. He received a Ph.D. in Economics from Syracuse University in 1973.

Jeffrey Burkhardt, Professor of Agriculture and Natural Resource Ethics and Policy, Food and Resource Economics Department (FRED), Institute of Food and Agricultural Sciences, University of Florida. He received his Ph.D. in Philosophy with a graduate minor in Economics from Florida State University in 1979, and joined the faculty of the University of Florida in 1985. He currently teaches courses on Agriculture and Natural Resource Ethics, Science Ethics, and the Philosophy of Economics.

Lawrence Busch, University Distinguished Professor of Sociology and Director of the Institute for Food and Agricultural Standards at Michigan State University. His interests include food and agricultural standards, food safety policy, biotechnology policy agricultural science and technology policy, higher education in agriculture, and public participation in the policy process.

Kenneth David, Ph.D., M.B.A. is Associate Professor of Organizational Anthropology and Trans-Cultural Management at Michigan State University. He received his Ph.D. from the University of Chicago and his M.B.A. from Michigan State University. His organizational Anthropology research in France, Holland, India, South Korea, Sir Lanka and the United States, focuses on such inter-organizational relationships as acquisitions, joint ventures, and engineering outsourcing design projects.

George Gaskell, Professor of Social Psychology, Pro-Director of the London School of Economics and Political Science. He is Associated Director of BIOS, the Centre for the study of Bioscience, Biomedicine, Biotechnology and Society at the LSE. From a background in social psychology, his research focuses on science, technology and society, in particular the issues of risk and trust, how social values influence people's views about technological innovation, and the governance of science and technology.

Mickey Gjerris, Assistant Professor, Danish Centre for Bioethics and Risk Assessment (CeBRA), University of Copenhagen. His research falls mainly within the areas of bio- and nanotechnology, especially focusing on the ethical issues surrounding the use of animals and the novel technologies. This research is embedded in the context of ethics of nature and religious philosophy and has as its point of departure the philosophical tradition phenomenology.

Hans Geerlings, Shell Global Solutions International B.V. and Delft University of Technology. He holds a Ph.D. in Physics from the University of Amsterdam. He does exploratory research – working as a Principal Researcher at the Shell Research and Technology Center and as a Visiting Professor in the Faculty of Applied Sciences at Delft University of Technology. His research interests include hydrogen storage in metal and complex hydrides, as well as carbon dioxide sequestration through mineralization.

John R. Lloyd, Department of Mechanical Engineering, Michigan State University is a University Distinguished Professor of Mechanical Engineering. His research program includes the emerging areas o energy transport at the nano and molecular length scales, which will have application in developing such diverse areas as thermal energy transport in Agrifood systems, thermoelectric devices, fuel cells, and energy efficiency in phase change heat transport in structured, micro, nano, and molecular scale thin film coatings on particles such as seeds and agri-elements.

Alan McHughen, Department of Botany and Plant Sciences, University of California-Riverside. After earning his doctorate at Oxford University, he worked at Yale and the University of Saskatchewan before joining the University of California, Riverside. A molecular geneticist with an interest in applying biotechnology for sustainable agriculture and safe food production, he served on recent National Academy of Science, Institute of Medicine and OECD panels investigating the environmental and health effects of genetically engineered plants and foods.

Philip Mancnaghten, Phil Macnaghten, Professor of Geography and Director, Institute of Hazard and Risk Research (IHRR), Durham University. He holds a degree in Psychology (1987, Southampton) and a Ph.D. in Social Psychology (1991, Exeter). He studies the cultural dimensions of technology and innovation policy and their intersection with the environment and everyday practice.

Margaret Mellon, Union of Concerned Scientists, Washington, DC. She came to the Union of Concerned Scientists (UCS) in 1993 to direct a new program on agriculture. The program promotes a transition to sustainable agriculture and currently has two main focuses: critically evaluating the use of biotechnology in plant and animal agriculture and assessing animal agriculture's contribution to the rise of antibiotic-resistant diseases in people. Trained as a scientist and lawyer, she received both her Ph.D. and J.D degrees from the University of Virginia.

Susanna Priest, Professor, Hank Greenspun School of Journalism & Media Studies, University of Nevada, Las Vegas. Her research and teaching focus on communicating science technology, environment and health; public perceptions of policy issues and public opinion formation, especially for these areas; mass media's changing role in society; media theory and research methods.

David Sparling, Associate Dean, Research and Graduate Studies, College of Management and Economies, University of Guelph. He was formerly and Associate Professor in the Food, Agriculture and Resource Economics at University of Guelph. He also farmed for twenty years near Cambridge, Ontario and has been president of an agribusiness insurance company and a biotechnology start-up. He is also a Senior Associate at the University of Melbourne. His teaching and research interests are in the areas of operations and supply chain management and commercialization of new technologies including a study of biotechnology IPOs in Australia and Canada.

Paul B. Thompson, Professor of Philosophy, Agriculture Economics and Community, Agriculture, Recreation and Resource Studies and W. K. Kellogg Chair in Agricultural, Food and Community Ethics, Michigan State University. He formerly held positions in philosophy at Texas A&M University and Purdue University. His research has centered on ethical and philosophical questions associated with agriculture and food, and especially concerning the guidance and development of agricultural technoscience.

Any K. Wolf, Oak Ridge National Laboratory, Oak Ridge, Tennessee. She leads the Society-Technology Interactions Group within the Environmental Sciences Division. Much of her research centers on the processes by which society makes and implements decisions about controversial and complex science, technology, and environmental issues. In addition, her work focuses on linkages between the conduct of science and the use of science in decision making. She received a Masters degree in Regional Planning and a doctorate in Anthropology from the University of Pennsylvania.

Analytic Introduction

Socio-Technical Analysis of those Concerned with Emerging Technology, Engagement, and Governance

Kenneth David

In a nutshell: our[1] audiences and our core objective

The emerging field of nanotechnology attracts antagonists (proponents and opponents), analysts from various disciplines, and a set of stakeholders: scientists, engineers, technology developers, research administrators, policymakers, standards-setting and regulatory agencies, non-governmental organizations (NGOs) and business executives, consumers, and citizens. This introduction addresses these diverse audiences with a communication strategy I learned from Ted Koppel, formerly of ABC News: Do not assume that the audience is ignorant. Also do not assume that the audience is sufficiently informed.

What can these antagonists, analysts, and stakeholders learn from the international controversy over the use of biotechnology involving recombinant DNA techniques in agriculture to produce "genetically modified organisms"? Biotechnology faced obstacles both in governance (standards-setting and regulatory agencies) and in social acceptance by buyers in the supply chain and by the public. The multinational agriculture and biotechnology company Monsanto, for example, withdrew its modified potatoes after they were rejected by two major buyers: Frito Lay and McDonald's. Monsanto's genetically modified (GM) corn seed was passed by governing agencies and accepted by farmers but faced much resistance from the final buyer—the consumer.

So can lessons from biotechnology be effectively modified and applied to the much broader field of technologies collectively called "nanotechnology"?

The objective of this volume is to collect analyses with different perspectives but with the common goal of providing lessons from biotechnology for nanotechnology. In it, the contributors present issues that occurred during the development of biotechnology and effective practices for responding to these issues that provide partial orientation for the development of nanotechnology. Each new technology (such as nuclear energy and biotechnology) poses particular challenges and hazards as well as benefits. There are environmental, social, and ethical impacts as well as technical and economic impacts. Formal standards, codes, and effective practices developed to deal with the impacts of earlier technologies cannot be applied wholesale to another new technology. Modifications in standards and practices must be made. In this volume, we study historical practices in order to modify them as necessary to meet the current set of impacts.

In Chapter 13, Busch and Lloyd succinctly set out a more specific set of questions: "Will the new nanotechnologies encounter the same or similar resistance? Are there lessons that we can learn by examining the failures and successes of agricultural biotechnologies? Can we shape the new nanotechnologies as well as respond to the concerns of critics and skeptics? What lessons can we learn from the experiences with the agricultural biotechnologies that will help us avoid the same result with the design of nanotechnological products and processes? What actions on the part of companies and governments might ensure the rapid and satisfactory resolution of concerns about nanotechnologies? What actions are likely to enhance public support for the promises that these new technologies bring? And what actions are likely to diminish that support?"

Finally, the overall intention of this volume is to make a collection of diverse perspectives on the topic of emerging technology. The objective of this introduction, then, is to highlight the contribution of this volume: to recognize contending perspectives with which various stakeholders or analysts deal with a controversial new technology.

This introductory chapter begins with a section on nano-benefits, nano-issues, nano-fears, and reactions, continues with a section on the objectives of this volume, and concludes with a "roadmap" to this volume.

Nano-benefits, nano-issues, nano-fears, and reactions

"Nanotechnology" relates to the science and engineering of materials and devices with dimensions between 1 and 100 nanometers. One nanometer is one billionth of a meter (approximately 80 000 times smaller than a human hair).

New technologies always stir controversy over hazards and benefits, and nanotechnology is no exception. It creates hope and excitement about possible breakthroughs for solving some of society's pressing problems. It raises social, ethical, and legal issues, and it also raises fears—angst that "nature" becomes partially constructed by humans.

Nano-benefits

Why did the US Government invest more than $1 billion in nanotechnologies in 2005? Possible nano-benefits are no secret. Berube's *Nano-Hype* (2006) amply records the extraordinary, "hyperbolic" claims made for applications of nanotechnology and Mehta (2004) provides a selection of applications expected to emerge from advances in nanoscience:

Environmental
- Remediation of contaminated soil and water
- Reduction in the use of raw materials through improvements in manufacturing
- Rebuilding the stratospheric ozone layer with the assistance of nanobots.

Medical
- Improvements in the delivery of drugs
- Development of techniques in nanosurgery
- Mechanisms to repair defective DNA
- Improved diagnostic procedures.

Electronic
- Development of molecular circuit boards
- Improved storage of data
- Development of molecular computers.

Materials
- Industrially valuable fibers with increased strength
- Replication of valuable products (e.g. food, diamonds)
- Improvements in the quality and reliability of metals and plastics
- Manufacture of "smart" materials.

The notion of a single "nanotechnology" is erroneous. In reality we are dealing with many nanotechnologies with multiple functions and multiple directions.

Nanotechnology is expected to foster a multi-billion dollar business with "nanomaterials" playing a prominent role. Among nanomaterials are polymer nanocomposites. Polymer nanocomposites have emerged as a new class of materials that has attracted the attention of researchers and industry across the world. Polymer nanocomposites are predicted to find multiple applications in various sectors of the economy, such as packaging, coatings, consumer goods, automotive, construction materials, structural materials and even homeland security. (Mohanty, 2006)

The promise of nano-benefits has also become part of popular culture.

Are NT devices small, but stable and helpful? Picture IBM's 2005 on-demand Business Help Desk commercial. A truck screeches to a halt in front of a desk in the middle of a deserted road. When the driver asks why she is there, the professionally suited woman tells the driver that she is at the Help Desk and that they are lost. The driver asks how she knows. She replies that the boxes have Radio Frequency Identification [RFID] tracking chips. The driver's buddy then dryly remarks, "Maybe the boxes should drive." (Wolfe *et al.*, 2006)

This scenario suggests that humans can now attain a degree of information precision never previously attained, as well as the possibility of a new organizational structure—a very flat organization capable of controlling and coordinating activities.

In short, potential nano-benefits have been forecast in many directions.

Social, environmental, biomedical, legal, and ethical nano-issues

The multiplicity of concerns raised by nanotechnologies matches the multiplicity of promises. Issues can be discerned by the following list of topics raised by experts attending a risk analysis conference in Brussels in 2004 (European Commission, 2004).

- Security problems
- Moving the nanoscience and technology debate forward towards short-term impacts, long-term uncertainty and the social constitution
- Mapping out nano-risks: considerations on possible toxicity
- Engineered nanomaterials and risks
- Nanotechnology—from the insurer's perspective
- Emerging concepts in nanoparticle toxicology
- Risks and ethical challenges of nanotechnology in healthcare.

What are the social, legal, and ethical[2] impacts of a controversial set of technologies? What issues stem from these impacts? Are there unambiguous answers to these issues?

Privacy

Invasion of privacy is a good example. Loyalty cards that include an RFID chip to identify customers and their purchasing preferences and facilitate micro-marketing to the customer are ethically questionable. So are "smart carts," shopping carts using scanning devices based on RFIDs. You walk through a supermarket. Each time you place an item in the smart cart, it is scanned. Then you approach the exit and find out that the cart has already read the credit card in your wallet. These perceived threats to privacy have already stirred protest by a group called CASPIAN (Consumers Against Supermarket Privacy Invasion and Numbering, www.nocards.org/).

In China, individual cows are already tracked via implanted RFIDs so that the incidence of bovine spongiform encephalopathy (BSE) can be revealed and countered (*MeatNews*, 2007).[3] To my knowledge, a bovine advocate has yet to appear to speak for the cows and against bovine privacy invasion. Cow producers, however, are another story, for tracing the origin of cows and tracking the progress from pasture to dinner table is perceived as violating the producers' right to privacy.

These examples show that there is no single ethical standard easily applied universally on the issue of privacy.

Hazard

Another issue is pure hazard. Medical researchers at the University of Michigan have already developed nano-scale devices that selectively destroy certain cancer cells. These devices are not ready for use, however, because they pierce holes through cell walls, leaving the cells vulnerable to infection. Insurance companies such as Swiss Reinsurance Company have done extensive work to anticipate corporate liability (and thus their own payouts) in the areas of environmental and biological hazards. Nano-risk, just like nano-applications, takes many forms.

> Coated nanoparticles can be extremely mobile in the environment. Once airborne, they can drift on more or less endlessly, since they—unlike larger particles—do not settle on surfaces, but are only stopped when, for example, they are inhaled or their dissemination is limited in some other way. On land, in the earth, and in the water, the same holds true. The smallest particles are washed through various earth strata and spread unhindered in a liquid medium, which means they pass easily through most filtering methods currently in use. (Swiss Re, 2004, p. 4)

Other sections of this report on the biological impacts of nanoparticles includes such subtopics as "Inhalation of nanoparticles," "Particle absorption though the skin," and "Particle absorption via the alimentary canal."

For a good recent review of the environmental risks of nanotechnology, see Dunphy Guzmán *et al.* (2006).

In short, fears and concerns about nanotechnologies, just like the benefits anticipated for nanotechnologies, take many forms.

Resources for research on risk assessment

Are sufficient resources being allocated for risk assessment? Is progress in standards setting hindered because resources for risk assessment are insufficient? The supplement to the US President's 2006 budget recommends $1.05 billion for overall National Nanotechnology Initiative investments. Of this amount, only $82 million is budgeted for societal dimensions:

- $38.5 million for environmental, health, and safety R&D
- $42.6 million for education and ethical, legal and other social issues.

Recent official reports find these allocations inadequate.

> Andrew Maynard, chief science advisor for the Wilson Center's Project on Emerging Nanotechnologies, said his analysis found the government spent only about $11 million in 2005. At the hearing, Maynard called for at least $100 million over the next two years for "targeted risk research." (von Bubnoff, 2006)

The National Nanotechnology Initiative, created by the Clinton administration in 2000, coordinates the many federal agencies that fund nanotechnology research. In 2003, Congress mandated that the National Research Council, an arm of the National Academies, conduct triennial reviews of the initiative. This council reported that research on how nanotechnology affects human health and the environment must be expanded.

> More safety research was also one of the recommendations of the National Research Council's triennial assessment of the NNI. The Congressionally mandated report, released on September 25, calls the results of safety studies "inconclusive," and states that there are too few studies that address the effects of nanomaterials *in vitro* and *in vivo*. (von Bubnoff, 2006)

Philosophical issues: the ontological angst of nanotechnology

Anthropologists noted long ago (e.g. Malinowski, 1922) the difference a society ascribes to a technology considered just adequate to deal with its intended usage and a technology considered dubious at best of being capable of coping with its intended function. In certain island cultures, for example, lagoon-worthy canoes, can be built by anyone—they require no ritual. Sea-going canoes, on the other hand, are produced by specific, skilled carpenters, are ritually decorated, and then certified by holy men (Figure 1.1). Ritualization is necessary when humans are fearful.

As technology advances, fears may subside. Alfred Nordmann, a philosopher of technology and society, has analyzed the roots of our fears around the progression of technology in society. Centuries ago, nature was uncanny, unpredictable, and sometimes dangerous (e.g. the black plague). Progressively, human science, at least as we know it in the West, technologized nature (Nordmann, 2005). That is, scientists and technologists gradually reduced the uncertainties of specific bits of nature and thus tamed bits of nature technologically. In the eighteenth century, for example, Benjamin Franklin showed the connection between lightning in the heavens and what was then called "scintilla"—the sparkling specks produced when

Figure 1.1 Sea-going canoes with elaborate prows from Kiriwina Islands (formerly known as the Trobriand Islands), Papua New Guinea (galenfrysinger.com 2006)

wool was rubbed the right way. Increased knowledge reduced onto-logical angst regarding nature. From the beginnings of agriculture in Neolithic times to genetically modified foods in current times, humans have been attempting to tame nature and cultivate what we consider socially necessary. Now, with the exploration of nanotech-nological frontiers, we perceive that we are messing around with the basic building blocks of nature, such as a nano-ring (Figure 1.2).

Are we entering a realm of the unknown again, this time inhabited by an uncontrollable pseudoscientific reality of uncontrollable nanobots—fears of self-replicating self-organizing nanomachines as portrayed in Michael Crichton's novel *Prey*? These fears, whether rational or farci-cal, elevate the possibility of a new uncanny nature of nature to a very real status—have we created a new uncontrollable nature and thus cre-ated a new ontological angst? In this volume, for example, in Chapter 4 Margaret Mellon states that nanotechnology may raise the "same con-cerns about the meaning of being human and our relationship to nature" (p. 85) as did biotechnology. In his book *Nano-Hype*, Berube contrasts two interpretations of nanotechnology:

> Is the technology only about chemosynthesis, catalysis on the nanoscale? Or is the technology about nanobots working together? If the former interpretation is accurate, then we need to examine the consequences of nanoparticles in terms of its interaction with the environment and its impact on life and world values. If the latter

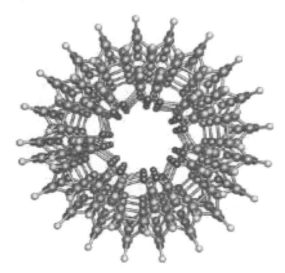

Figure 1.2 Nano-ring

interpretation is accurate, then we may need to consider whether a world with nanobots doing our bidding is such a good idea. Or maybe we are approaching something between the two interpretations. (Berube, 2006, p. 21)

This split between nano-scale chemosynthesis and nano-scale mechanical manufacturing is important in the dialogue between proponents and opponents of nanotechnologies. Further, the next two sections here—on marketing, de-marketing and counter-marketing of an emerging technology and on controversy and hyper-controversy among proponents and opponents—lead us directly to the definition of the objectives of this volume and the contending perspectives presented in this volume.

Marketing, de-marketing, and counter-marketing of an emerging technology

Even before the widespread mass marketing of nano-products has taken place, we can still distinguish processes of marketing, de-marketing, and counter-marketing of this emerging technology. A market in question is government funding of research.

On the "pro" side, scientists, whether in university laboratories or government laboratories such as Oak Ridge National Laboratory, have predominantly applied for (marketed) the chemosynthesis direction—the safe side of nanotechnology, and government funding predominantly favors chemosynthesis research and development.

Opponents, including NGOs such as ETC and Zac Goldsmith, the British environmentalist and editor of *The Ecologist* magazine, de-market nanotechnology by emphasizing the hazards of the nano-scale manufacturing side—the more frightening side of nanotechnology. In science fiction, Crichton's *Prey* is the latest in a series of popular representations that are perceived as opposition to contemporary scientific advances. People have long recognized reactions in the media against new technology (think of Charlie Chaplin rebelling against the machines in *Modern Times*). But how frequently are impacts tangibly demonstrated? I've been told by a public health policy administrator, for example, that although the human transplant industry has come a long way in modern medical miracles, the extreme controversy surrounding it, the media, and public fear are very hard factors to overcome. Every year, when the movie *Coma* is run on TV, national donation rates plummet for approximately 6–8 weeks.

Other proponents such as the Center for Responsible Nanotechnology (CRN) and various business leaders such as the NanoBusiness Alliance are counter-marketers. They undercut arguments made by nanotech opponents. Chris Phoenix of CRN spoke at our conference and delimited the field in this manner. He attacks Eric Drexler's utopian vision of "Engines of Creation," that is, self-replicating, molecular nanotechnologies. This argument thus questions some threats as perceived by the public. He further suggested that "education is needed to combat mis-education and misrepresentations of technology and ridiculous fears."

Reactions to an emerging technology: types of adversarial action

Reactions to the advent of nanotechnology are not tame. The ETC Group (Erosion, Technology, and Concentration) has called for a moratorium on commercialization of products until there is more adequate coverage of safety concerns. They maintain that at present there is inadequate understanding of nanotechnological risks and that effective practices for handling and using nanoparticles have not been established (ETC Group, 2003).

CASPIAN hosts a website (www.spychips.com) that attacks practices such as the inclusion of RFID chips in products by the German supermarket chain Metro. They point out that customers are not aware that RFID chips embedded in their Metro loyalty cards could identify and track their purchases (CASPIAN, 2004).

I suggest that nanotechnologies are facing something more than mild controversy. Nanotechnologies are likely to come against three types of adversarial situations—dispute, controversy, and ultra-controversy—with accompanying modes of dialogue and modes of resolution.

Dispute

A dispute involves a discrete contested issue. Dialogue is possible between parties to a dispute. Dialogue may require legal process to resolve the dispute. Resolution is possible within the existing rules of the game. Each disputant tries to frame the issue according to rules that favor his or her position. The outcome does not necessarily change the rules of the game.

Controversy

A controversy involves more ambiguous and complex issues. Dialogue is established only with difficulty; mediators may be necessary. Opponents are not initially willing to talk to each other but they may come to recognize that a common ground exists. Opponents do not clearly understand each others' perspectives. Resolution is a protracted, iterative process. Education of opponents to understand both sides of the controversy is necessary in order to move towards resolution. Opponents may eventually show a willingness to consider each others' positions seriously.

Ultra-controversy

Various features and trends define this adversarial situation.

First, an ultra-controversy does not appear to involve discrete issues. An antagonist can bundle together a series of controversial issues such as globalization, capitalism, government repression, biotechnology, and nanotechnology. "Top hoppers" who appear at global meetings such as the World Trade Organization, the G8, etc. present arguments vilifying a bundled set of issues. Debundling issues is typically unsuccessful.

Second, mutually exclusive perspectives exist; antagonists polarize themselves into extreme positions. There is no simple binary contrast encompassing all positions; rather there is a means/extremes type of contrast. This is expressed by Wolfe and Bjornstad in Chapter 8 with their trichotomy of opponency positions: Absolute Rejection ... Everything in Between ... Absolute Acceptance. Extreme antagonists either absolutely reject or absolutely accept the emerging technology. They appear to be speaking a different language. Antagonists do not necessarily recognize each others' right to address the topic. Opponents to technology, for example, may "demonize" the proponents. On the other hand, staunch proponents to the technology may "idiotize" the opponents.

Third, over time, there has been an increasing international political sensitization due to a series of previous "controversial" technological issues:

1. Nuclear energy production versus nuclear weapons grade production and nuclear proliferation—post World War II.
2. Cloning to reduce adverse traits versus cloning as racist eugenics leading to the production of a limited gene pool.

3. Improved computer-aided communication versus invasion of privacy of computer users.
4. Globalization of capitalism as a source of unprecedented wealth versus globalization of capitalism as the root of inequality and hyper-competition. Proponents focus on new tools and the potentialities they bring into existence—the Internet and other forms of communication, increasing access to information sharing, and increased access to capital in its many forms. Opponents are generally quite politicized and tend to attack the highly developed capitalistic economy steered by multinational corporations whose operations foster difficult aspects of globalization.
5. Biogenetic agriculture as improved production versus "Frankenfood" image of GM foods.

Fourth, as George Gaskell indicates in Chapter 12, this series of events resulted in a qualitative change: a questioning of scientific and technological authority. With the advent of nuclear power, computers, and modern biotechnology or the life sciences, the three strategic technologies of the post World War II decades, a cleavage between science, technology and society has appeared. Increasingly, sections of the European public have questioned whether the good life, as defined by science and technology, is actually what they, the public, aspire to. This cleavage turned into open conflict in Europe over GM crops and food; a controversy that became emblematic of the questioning of scientific expertise and of the established procedures of risk governance.

Fifth, there is sharper and quicker communication of protest events both in public media and in internet-based communications such as blogs. Control of the mass media by corporate interests does not, therefore, totally block communication of events and major publicity is guaranteed because of intense reporting of the series of anti-globalization demonstrations (Seattle; Genoa etc. demonstrations against World Trade Organization, World Bank, OECD nations meetings).

Regarding mode of dialogue, an "ultra-controversy" is marked by negative dialogue; mutual denigration of the opposite position ("demonization" of the technical advocates; "idiotization" of the anti-technical advocates) can occur. Inflammatory statements are made with no expectation that antagonists shall seek common ground. Mode of resolution of ultra-controversy is not yet known.

Summary

This section addressed three kinds of complexity regarding the advent of nanotechnologies. First, we ascertained that nano-benefits, nano-issues, and nano-fears all exist. Second, we discussed the three forms of marketing that are a reaction to nanotechnologies. Nanotechnologies incur negative de-marketing messages by opponents. They also receive positive (or, according to Berube, hyperbolic) marketing messages from proponents. Counter-marketing, that is, countering the negative messages, also occurs. Third, nanotechnologies are likely to face all three forms of adversarial situations: disputes, controversies, and hyper-controversies. Further, regarding the discussion of types of adversarial action, understanding the spin about nanotechnologies requires attention to three types of adversarial action. Dialogue is possible between disputants. It may be established with some difficulty between protagonists (proponents and opponents) to a controversy, but it should not be expected of participants on the ultra-controversy mode of adversarial action. It is not likely, therefore, that any form of social dialogue will be developed that will satisfy all stakeholders and all analysts of biotechnology and nanotechnology.

Given these complexities we hold that no single, overarching theoretical framework is capable of properly addressing these topics. How shall we address these topics? The next section clarifies our intentions in this volume.

Objectives of this volume

This volume is an intentional collection of diverse perspectives on whether and, if so, how we can learn from the international controversy over biotechnology as we now face the onset of nanotechnologies. (Those who want a detailed definition of genetic engineering, the key process of biotechnology, can turn to Alan McHughen's Primer on Genetic Engineering in Appendix I).

The authors whose work is collected here met at the First International Institute for Food and Agricultural Studies (IFAS) Conference on Nanotechnology that convened at Michigan State University, East Lansing, Michigan on October 26 and 27, 2005. The Conference was titled "What Can Nano Learn from Bio? Lessons from the Debate over Agrifood Biotechnology and GMOs." We met in public conference mode for 1.5 days and then in workshop mode for another 1.5 days.

The editors of this volume share certain working principles. We start with the view that nano-benefits, nano-issues, and nano-fears all exist. No overarching theoretical framework is capable of properly addressing all these topics. We shall not present one totally unified, coordinated theory. We are not lobbying for one particular perspective.

We do, however, intend to limit the presentation in one particular way. To study a controversial technology we distinguished degrees of adversarial social agitation: disputes, controversies, and ultra-controversy. Our criterion for inclusion of works in this volume is that we are dealing with presentations of opponency and proponency of a controversial issue, not the more limited contestations by parties to a dispute and not the more extreme presentations we have called ultra-controversy. Rather, we intend to make these topics (nano-benefits, nano-issues, and nano-fears) more accessible by bringing together an ordered collection of perspectives representing diverse stakeholders in the onset of nanotechnologies and diverse analysts who have studied such controversial technologies as bio- and nanotechnologies.

More specifically, analysts may well be grouped into three disciplinary categories: philosophical and ethical reflections on STS (science, technology, and society), natural science analyses of STS, and social science analyses of STS. All three perspectives are represented here.

Further, there are a set of stakeholders in the emerging field of nanotechnology: scientists, engineers, technology developers, research administrators, policymakers, standards-setting and regulatory agencies, NGOs and business executives, consumers, and citizens. What can these stakeholders learn from the international controversy over biotechnology?

The authors were charged with presenting papers that covered a spectrum of perspectives on biotechnology controversies. They also were charged with discussing whether the controversies over biotechnology are helpful to provide guidelines for acceptance or rejection of processes used or devices produced by nanotechnologies. The results—the contributions to this volume—do not show a night and day distinction between the work of stakeholders and that of analysts. Stakeholders also analyze the situation; analysts have some stake in the situation.

Contending perspectives

Continuum of opponency and proponency

The earlier discussion of types of adversarial situations (dispute, controversy, and ultra-controversy) and types of marketing (marketing, de-marketing, and counter-marketing) can now be put to work. Figure 1.3 summarizes the contending perspectives represented in the volume. You will note that these contending perspectives do not

Continuum of opponency and proponency to biotechnology and to nanotechnologies:

Extreme opponency: hyper-controversial groups lump a variety of controversies together and reject all of them *Demonize opponents:* Top-hoppers; some NGOs	Opponency	Proponency	Extreme proponency: pursue progress because it can be done *Idiotize their opponents:* Some scientists; some venture capitalists

←HYPER-CONTROVERSY→   ←   CONTROVERSY   →   ←HYPER-CONTROVERSY→

No dialogue possible

Current situation:
Diversity and separation of stakeholders
Objective: dialogue among stakeholders

No dialogue possible

Opponency		Mediation		Proponency		
Public advocacy with opponency	Public advocacy with principled progress	Mediation for principled progress	Scientist plus mediator	Facilitating and implementing the technology	Support for the controversial technology	Scientist with some sense of caution
Watchdog; **de-marketing** of emerging, risky, technology; earnest opponency	Studied neutrality Questioning of values & principles to avoid unbridled scientific/ technical action	Activities to incite public acceptability and to facilitate public involvement in decision-making	Scientific progress with awareness of need for public acceptance	Implementing new technologies via business organization: via legal procedures; via patent procedures; via media	**Counter-marketing** of watchdog messages; making responsible scientific action apparent to selected audiences	Scientific progress with safeguards against undue risk **Marketing** of progress
	Equitable distribution of dissatisfaction; Mediation of scientific, technical, business resource allocating, standards-setting, regulatory, and public stakeholders					

Figure 1.3 Perspectives appearing in this book

map exactly with the contributions by individual authors. Individual authors espouse different perspectives on different key issues and some entries represent the perspective of speakers at our conference who have not contributed a chapter. Nevertheless, it is a useful point of departure to collect and arrange these perspectives.

Key relationships and issues: engagement, supply chain, governance, and resource allocation

The next step is to specify themes (key social relationships and issues) that were indeed addressed by the contributors to the volume.

Engagement

Engagement of the scientific/technical community concerning an emerging, controversial technology is a theme touched, directly or indirectly, by all the contributors. Engagement includes topics such as upstream engagement, democratic participation in dialogue, and prevalence of the "knowledge deficit" model, that is, one-sided, stratified communications from the scientific community to the public. In such engagement, communications are indeed mediated by the mass media (Priest, Chapter 11) and by citizen advocates and NGOs (Mellon, Chapter 4). Further, two authors (Burkhardt, Chapter 3 and Gjerris, Chapter 5) particularly question the advisability of one-sided communications between scientists and the public. McHughen's perspective (Chapter 2) is that of a natural scientist who is addressing natural scientists who did not pay enough attention to these issues during the biotechnology controversy. Geerlings and David (Chapter 10) discuss viable timing of engagement from the perspective of a natural scientist working with a social scientist.

Supply chain issues

A set of contributors discuss competitive and cooperative relationships in the supply chain that affect the development and commercialization of nanotechnology applications. Whether in academia or in business, the relationship between scientific and technology innovators on one hand and resource allocators is a key factor in the process of innovation. McHughen (Chapter 2), Sparling (Chapter 9), Geerlings and David (Chapter 10), and Busch and Lloyd (Chapter 13) present contrasting views regarding innovation in the supply chain from the points of view of natural scientists, social scientists, and management scholars.

Governance issues

Governance is the relationship between standards-setting and regulatory agencies on the one hand and technology innovating companies on the other hand. The construction of new realities in the form of standards and codes by standards-setting and regulatory agencies is discussed by Busch and Lloyd in Chapter 13.

The key themes addressed by the contributors to this volume are summarized in Figure 1.4 and Table 1.1.

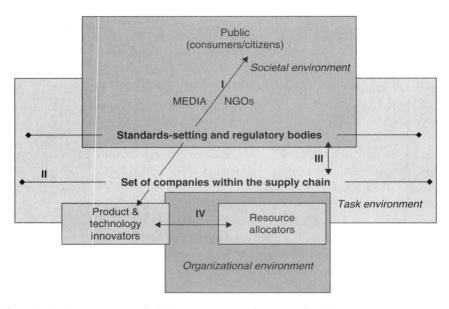

Figure 1.4   Engagement, supply chain, governance, and resource allocation

Table 1.1	Main relationships as identified in Figure 1.4
I	Relationships between the science/technical community and the public; communications are modified, augmented, and transformed both by mass media and by NGOs
II	Relationships among companies in a supply chain. Supply chain constraints impact on technological development
III	Relationships between standard-setting and regulatory organizations on the one hand and companies in the supply chain on the other hand
IV	Relationships among scientists, engineers, business managers, etc. in the organizational environment.

Roadmap to this volume

We now continue with a preview of offerings in this volume—including a brief description of each author to indicate the perspective that appears in their writings.

Following *Part 1 Analytic introduction*, three chapters present varying perspectives in *Part 2 Looking back to the biotechnology debate*. A natural scientist, a philosopher, and a dedicated advocate of public engagement bring diverse perspectives to this topic.

- *Alan McHughen* is a natural scientist who specializes in biotechnology. In Chapter 2 he takes the perspective of a natural scientist who considers both technical and non-technical obstacles to technological innovation. The fledgling nanotechnology community might learn from another recent technology, biotechnology. The technical and non-technical history of modern biotechnology, complete with missteps, is presented here, focusing on those aspects of greatest relevance to nanotechnology in the hope that the nanotechnology community might avoid or otherwise prepare to overcome these obstacles. In Appendix I, McHughen presents a short Primer on Genetic Engineering.
- *Jeffrey Burkhardt* is an agricultural economist and a philosopher of society and technology. He reviews in Chapter 3 the ethical considerations on the biotechnology debate: the nature of the technology, claims concerning health and environmental impacts, and disagreements over socio-economic impacts. This case study is a model for ethical debates likely regarding other emerging technologies. He argues that the scientific community (using the science model of rationality) has persistently failed to understand what critics are saying because they translate everything into consequences and trade-offs.
- *Margaret Mellon* is an advocate of public engagement from the Union of Concerned Scientists. In Chapter 4 she presents a view from the advocacy community, a strong call for restraint in implementing this emerging technology. According to Mellon, for many participants in the biotechnology debate the story is not primarily that of a technology that stumbled. She states that the public debate over biotechnology was productive in that it raised questions about how decisions are made about the technology: She calls for explicit questioning of how decisions are made about the technology and for more transparency in decision-making.

In *Part 3 Questioning the analogy (from bio to nano)*, four more chapters consider whether it is valid to take the extended public debate over biotechnology and GM foods as a source of lessons for issues regarding nanotechnology devices and processes yet to come. Do the public controversy and regulatory hurdles associated with agrifood biotechnology provide a useful model for anticipating similar hurdles in nanotechnology? Although many authors have appealed to this model in urging a cautionary and attentive attitude on the part of scientists, research administrators, and government regulators responsible for nanotechnology, it is important to ask whether the analogy is valid.

- *Mickey Gjerris* is a bioethicist and risk assessment scholar. In Chapter 5 he questions whether the core question of this volume, "What can nanotechnology learn from biotechnology?" is correctly posed: Is the question just a way of managing crises? He holds that it is too broadly focused and restates it as follows: "What can we as citizens, as members of societies, learn from the biotech experience about ethically scrutinizing new technologies in the best possible way?" He states lessons from the biotechnology debate as follows: first, forget the knowledge deficit model, second, avoid one-sided debate, and, third, enjoin scientists to listen to the public.
- *Philip Macnaghten* is a geographer and director of the Institute of Hazard and Risk Research who studies the embodied dimensions of people's experience in, and of, technology, the future, and the natural world. He suggests in Chapter 6 that directly learning lessons from the GM food controversy and applying them to nanotechnology is only partially right. Through empirical research with regulators and the public, the author examines the lessons to be learned from this experience, particularly in relation to the governance and regulatory responses to new and emerging nanotechnologies. In particular, he outlines the need for more textured, socially realistic analysis of the distinctive character of particular technologies, and greater recognition of the limitations of conventional models of risk assessment.
- *Paul Thompson* is a philosopher of technology and society. He also questions the bio to nano analogy but with a different analytic method. In Chapter 7 he offers 10 reasons to think that it is not valid, and then subjects each of them to a critical discussion. The result of this systematic comparison is that the analogy

between agrifood biotechnology and nanotechnology certainly needs to be qualified. Many of the points on which nanotechnology might be importantly different from biotechnology depend on what the developers of nanotechnology do from this point forward, while others apply more strongly to some probable applications of nanotechnology than to others.

- *Amy Wolfe*, an anthropologist, and *David Bjornstad*, an economic policy analyst, work with natural scientists at the Oak Ridge National Laboratory. In Chapter 8 Wolfe and Bjornstad go beyond that biotech/nanotech analogy to question what is or is not comparable (in terms of societal responses) across a larger suite of emerging technologies, and then suggest developing a societal response science. This societal response science would help provide the conceptual/theoretical basis for determining what is or is not comparable across an otherwise disparate, disconnected (or, not necessarily connected) set of studies.

Part 4 Areas of ambiguity in implementing an emerging technology presents organizational, supply chain, and media issues that bear on technological innovation and the introduction of potentially controversial technological devices.

- *David Sparling* is an agricultural economist business scholar. He holds that the impacts of biotechnology, first, offer an opportunity to anticipate challenges of nanotechnology and, second, foreshadow impacts of nanotechnology on business models, business operations, and the structure of industries adopting nanotechnology. While standard business strategy innovation is defined in terms of product, process, and target market innovation, Sparling adds organizational innovation. He traces both the stages of technical (scientific–technological–commercialization) innovation and corresponding organizational innovations necessary for implementation.
- *Hans Geerlings* is a principal research scientist at Shell Global Solutions and Professor in Applied Sciences, Technical University Delft and *Kenneth David* is an organizational anthropologist. In Chapter 10 they present two issues—engagement and translation—that relate to engagement among four parties: Scientists, resource allocators (in academia or in business), the public, and governing agencies. Timing of engagement involves optimizing timing of reliability of risk assessment and engagement among the four

parties. Translation issues recognize that when a technology is emerging, multiple messages are communicated via various media of communication to diverse audiences.

● *Susanna Priest* is a media scholar. In Chapter 11 she considers two conflicting roles of the media in communicating scientific developments to the public: as public engagement and as market research. While some advocates of public engagement in the formation of technology policy seek ways to improve deliberative democracy, others are likely more concerned with heading off— or at least identifying—"problems" with public acceptance.

Finally, in *Part 5 Looking forward to the nano situation*, more specific lessons are drawn from the biotechnology debate for the onset of nanotechnologies in these chapters. Regulatory, legal, social and engineering perspectives appear here.

● *George Gaskell* is a professor of social psychology. The European experience of modern biotechnology provides a number of lessons with emerging technological innovations such as nanotechnology. There are the dangers of "group think" centered on hubris and hype among the promoters of the technology. More specific lessons include the need to anticipate the consequences of, first, signing up to international agreements; second, ignoring and/or dismissing the repeated warning signals of concerned public opinion; third, adhering rigidly to a narrow "sound science" approach to the assessment of risks and benefits; fourth, failing to appreciate that the hurdles to successful innovation go beyond regulation and the traditional definition of the market; fifth, assuming that science trumps all other consideration including social values; and, sixth, not recognizing the need to "pave the way" for innovations as they enter the public domain.

● *Lawrence Busch* is a professor of sociology and director of the Institute for Food and Agricultural Studies who studies social issues regarding food standards. *John R. Lloyd* is a mechanical engineering professor specializing in thermodynamics and nanotechnologies. In Chapter 13 they present a few succinct lessons and conclusions distilled from the preceding chapters, aimed primarily at an audience of practicing scientists and engineers. The authors suggest that although agricultural biotechnologies have enjoyed some successes, they have failed to live up to the promises and claims of the early 1980s. Some reasons for the many failures of agricultural biotech are highlighted, including regulatory issues and marketing. Nanotechnology is now traveling along the path of scientific innovation and public marketing. In order for nanotechnology to live up to its promises, the authors suggest five specific lessons from agricultural biotech that they illustrate with their case studies.

We append a further offering that updates the main set of contributions. The 2005 International Conference generated the text of this book. In 2006, we convened another international gathering, a workshop which was more participatory and more narrowly focused on one of the themes that emerged from the material in this book: standards. Five topics appear in Appendix II—Standards for nanotechnology workshop report: Timing and standards setting, Product versus process standards, International harmonization, Integration of operational standards, and Participation and transparency in standards-setting processes.

In addition, Appendix III lists acronyms for organizations bearing on emerging technologies, Appendix IV lists participants at the 2005 Bio to Nano conference, and Appendix V lists participants at the 2006 Standards workshop.

Conclusion

This volume presents an intentional collection of diverse perspectives: natural science, social science/organizational studies, and philosophical/ethical studies. The collection is intentional in that we recognize that newcomers to this discussion experience a certain intellectual vertigo. Our collection may help reduce this vertigo by noting, first, that natural sciences indicate convergences of disciplines that were previously separate, and, second, that social sciences show continued fragmentation (aka balkanization) of disciplinary studies what should be brought together, and, third, that both philosophical and governance perspectives include strong positions of proponency and opponency.

- *Natural science of science and technology: disciplinary convergence*—In this perspective, nanotechnology refers to a convergence of enquiry by scientists from a variety of disciplines.

Nanotechnology borrows liberally from condensed matter physics, engineering, molecular biology, and large swaths of chemistry. Researchers who once called themselves materials scientists or organic chemists have transmuted into nanotechnologists.

- *Social science of science and technology: disciplinary separations*—A key barrier studying acceptable technology and to effectively using this knowledge is that researchers of social acceptability are balkanized along disciplinary and subject matter lines. Disciplinary boundaries of anthropology, sociology, economics, history, and philosophy tend to separate these researchers. Social scientists who have studied chemical technology and those who study biotechnology seldom compare notes.

- *Philosophy of science and technology: disagreements regarding ethical and analogic arguments*—Strong principled claims and arguments from both defenders and critics of emerging technologies. Arguments exist both for and against using biotechnology as an overall analogy for the practice of nanotechnologies—but noting the analogy possibly useful for policymakers struggling to handle the emerging set of technologies.

- *Governance of science and technology: proponency and opponency*—Regarding the regulation of bio/nanotechnologies there are contrasting positions. Proponents hold that regulation of nanotechnologies can be done well under existing codes and processes. Opponents (advocacy groups and NGOs) allege that current regulations are not sufficiently elastic to address the unique and novel risks to people and the environment posed by nano-particles. They propose new regulation to deal with the broad social, health, environmental, and economic concerns of technologies converging at the nano-scale.

Faced with such an array, it is not appropriate to attempt a forced integration of these arguments but rather to present a set of questions that arose during and after our 2005 conference—questions that conference participants and the Michigan State University NIRT research group consider priority questions that require further attention.

Governance

How will these new technologies be governed? What changes, if any, will be needed in government (local, state, national, international) regulation in order to inspire confidence in the use of these technologies as well as to avoid undesired impacts? What role(s) will the private sector need to play in governing these technologies, with respect to standards development, certification of products and processes, etc.? What are the implications for worker/consumer health and safety? How can some of these changes be foreseen so as to develop new nanotechnologies with this in mind?

Governance and supply chain activity
What strategic and ethical issues concerning standards and regulation should be addressed? What are the imperatives and limits of corporate social responsibility? Do power relations between supply chain captains and subordinate suppliers in the food industry impose significant standards-setting and "regulatory" action on the subordinate suppliers? Do major supply chain captains have significantly different regimes of action towards the subordinate suppliers?

Engagement/participation
To what degree can/should the public participate in decisions about nanotechnologies? What forms of participation might be most effective? How can cooperative extension help in building an effective dialogue with the public on nanotechnologies, especially with respect to the food and agricultural sector, but also with respect to broader environmental issues? What factors about new technologies or the way that they are developed and introduced tend to promote public acceptance, and what factors tend to provoke resistance? In addition to agrifood biotechnology and GMOs, other studies on acceptance and rejection of technology are beginning to lay the basis for a social science of acceptable technology.

Social/technical interface
What are the likely economic, social, ethical, and legal opportunities for and barriers to widespread adoption of various nanotechnologies for various participants in the supply chain from input supply through to final consumption? How will these be distributed among persons, families, regions, nations, income groups, etc.? How might standards for quantifying and validating information (e.g. traceability through use of nanosensors) facilitate or reduce adoption?

The contributions collected here discuss issues that occurred during the development of biotechnology and (in-)effective practices for responding to these issues. We suggest that these studies provide partial orientation for the development of nanotechnology. Forgetting history and repeating its ills is not an option. On the other hand, while weaving through exaggerated promises (e.g. nano-hype), ignoring possible contributions of the diverse set of nanotechnologies is also not an option. This volume provides a foundation for more constructive consideration and more effective practices to guide the development of nanotechnology.

Endnotes

1. We are the Agrifood Nanotechnology Research Group, a multidisciplinary research group centered at Michigan State University. Our project is funded by a US National Science Foundation NIRT grant. We have four main objectives: The first is to determine what lessons from the experience with public reactions to biotechnologies will be relevant to developing nanotechnologies. The second is to determine what kinds of social and ethical issues might be raised by the turn to agricultural nanotechnologies. The third is to determine what kinds of standards (e.g. technical standards, food safety, environmental, or marketing standards) will need to be developed in commercializing agrifood nanotechnology. The last is to examine social, ethical, and economic problems that might be encountered in developing these standards.
2. The systematic study of morality is a branch of philosophy called ethics. Ethics seeks to address questions such as how one ought to behave in a specific situation ("applied ethics"), how one can justify a moral position ("normative ethics"), and how one should understand the fundamental nature of ethics or morality itself, including whether it has any objective justification ("meta-ethics").
3. Radio frequency identification (RFID) technology for beef traceability has been launched in China to guarantee food safety from farm to table. The system, developed jointly by China Agricultural University and China Tagtrace Tech Ltd, is being tested among several leading beef integrators in Beijing, Shaanxi, and Liaoning, and is expected to be applied throughout China in the near future. The first batch of beef products under the system has been available

in Lotus, a supermarket chain within Thailand-based Charoen Pokphand Group in Beijing. Buyers can check information related to the products' quality and safety, such as the specific source of the beef, the animal's breed and age, as well as feed the animal was fed and its disease history, at the market, or by mobile phone or logging onto www.safebeef.cn. A traceability system is planned for more animal products in China, as a law on animal husbandry that requires strict tracing was recently passed in the country (www.meatnews.com).

References

Berube, D. M. (2006). *Nano-Hype: The Truth Behind the Nanotechnology Buzz*. Prometheus Books, p. 21.

Dunphy Guzmán, K. A., Taylor, M. R., and Banfield, J. F. (2006). Environmental risks of nanotechnology: National Nanotechnology Initiative Funding, 2000–2004. *Environ Sci Technol* **40**, 1401–1407.

ETC Group (2003). *No Small Matter II: The Case for a Global Moratorium. Size Matters!* ETC Group, Ottawa, Ontario, Canada.

European Commission (2004). *Nanotechnologies: A Preliminary Risk Analysis on the Basis of a Workshop Organized in Brussels on 1–2 March 2004 by the Health and Consumer Protection Directorate General of the European Commission*.

Malinowski, B. (1922). *Argonauts of the Western Pacific*. Routledge.

Mehta, M. D. (2004). From biotechnology to nanotechnology: What can we learn from earlier technologies? *Bull Sci Technol Soc* **24**, 34–39.

Mohanty, A. K. (2006). Syllabus Packaging 891: Multifunctional Nanomaterials: Course Description, Fall 2006. Michigan State University.

Nordmann, A. (2005). Technology naturalized, Plenary speaker at the Society for Technology and Philosophy conference at University of Delft, 2005. http://spt.org/

Wolfe, A. K., David, K., and Sherry, J. (2006). It depends on where you sit: Anthropologists' involvement with nanotechnology in government, university, and industry settings. In: Stone, J. V. and Wolfe, A. K., eds. *Nanotechnology in Society: Atlas in Wonderland? Practicing Anthropology* **28**, 2006.

Internet references

CASPIAN (2004). German RFID scandal: hidden devices, unkillable tags found in METRO future store. www.spychips.com/press-releases/german-scandal.html

MeatNews (2007). China develops traceability system on beef. www.meatnews.com/index.cfm?fuseaction=article&artNum=14224

Swiss Re (2004). *Nanotechnology: Small matter, many unknowns.* www.swissre.com/INTERNET/pwsfilpr.nsf/vwFilebyIDKEYLu/ULUR-5YNGET/$FILE/Publ04_Nanotech_en.pdf

von Bubnoff, Andreas (2006). EHS efforts caught in the crosstalk. *Small Times* November. www.smalltimes.com/news/display_news_story.cfm?Section=WireNews&Category=Home&NewsID=140334.

Looking Back to
the Bio Debate

PART

2

Learning from Mistakes: Missteps in Public Acceptance Issues with GMOs

Alan McHughen

What Can Nanotechnology Learn from Biotechnology?
ISBN: 978-012-373990-2

Introduction

A major problem in public discourse and debate over modern technologies is the lack of accurate information and a scientifically questionable foundation. This is certainly clear in the public debate surrounding agricultural biotechnology. The quality of the debate is generally poor, largely because the widespread ignorance of techniques used in ordinary food production precludes a logical comparator baseline. Lacking a solid understanding of the status quo baseline, people have no appropriate means to compare the risks and benefits of biotechnology.

The answers to questions taken from a survey by the Food Policy Institute (Hallman *et al.*, 2004) illustrate the level of knowledge of food and biotechnology in the US (Table 2.1).

These sample (but non-random) questions from the larger survey illustrate the poor level of popular knowledge of biotechnology. Particularly disturbing is the observation that the first six questions require binary answers—that is, simple "yes" or "no." That means a person who did not know the subject matter and simply guessed would have a 50% chance of getting the correct answer. Overall, if all respondents were completely ignorant and simply guessed at every answer, the results would be statistically close to 50%. But on these six questions, the correct answers range from 30% to 48%. In order to skew the results so far down and away from random guessing, a substantial portion of respondents must have thought they knew the correct answer, and so did not simply guess. But they were wrong in their "knowledge"! Although these select questions are not representative of the complete survey, they do serve to illustrate a major problem in the agbiotech debate—many people believe they know something about genetically modified organisms (GMOs) and

Table 2.1 Food Policy Institute survey	
	Survey results (% correct)
Are GM foods in US supermarkets?	48
Do ordinary tomatoes contain genes?	40
Would a tomato with a fish gene taste "fishy"?	42
If you ate a GM fruit, might it alter your genes?	45
Can animal genes be inserted into a plant?	30
Give an example of GM food on the market	79 tomatoes

therefore are confident in their answers to questions, but are wrong. This state of affairs is clearly worse than people recognizing and admitting they do not know much about GMOs and simply guessing answers to questions.

Another major problem is the use of terminology.

Problems with terminology

Some of the terms in common use are:

- Biotechnology
- "Modern" biotechnology
- Genetic engineering
- Transgenic
- rDNA (recombinant DNA)
- Genetic modification ("GMO").

Terminology is always a confounding factor in discussion and debate around any technical field. Those not "in the know" are disadvantaged, especially when opponents use a myriad of confusing and unknown terms. A particular problem arises when specific terms are used, which, although everyone claims to know what they mean, are used with differing definitions. When this happens, someone may claim to understand an opponent's point whilst completely misunderstanding their argument. This situation, which is common in the agbiotech debate, is actually worse than debates in which terms are totally unknown; standard definitions can then be sought. To complicate matters further, some "common" terms are not even defined consistently among the experts and regulators who use them in official capacities.

It is fair to assume that nanotechnology, a recent technological development replete with technical terminology and jargon, will face similar confusion and ignorance. The degree of such confusion will depend on the degree to which nanotechnology employs common words in its technical or semi-technical vocabulary. The fewer the better in general, as more confusion and ambiguity arises when common words are appropriated or usurped for technical nomenclature.

What is genetic modification/genetic engineering/biotechnology?

Biotechnology includes any of several techniques used to add, delete, or amend genetic information in a plant, animal, or microbe. It is used to make pharmaceuticals (insulin, dornase alpha, etc.), crops (Bt corn, disease-resistant papaya, etc.) and industrial compounds (specialty oils, plastics, etc.). More details are given in the Primer on Genetic Engineering (Appendix I).

History of biotechnology

If we define biotechnology as the application of science or technology to biological systems, as many do, then biotech started 10 000 years ago, when humans first decided to put down roots and modify their surroundings to meet their needs rather than chase the desired environment with the seasons, geography and so on. Human society has manipulated and altered living systems (i.e. applied biotechnology) ever since—particularly in the pursuit of food. Intentional selection of preferred seeds for planting, rather than for consumption, was an early manifestation of biotechnology. Even this early form of agricultural biotechnology left an unexpected human footprint: planting selected seeds meant displacing whatever species were occupying that land before. As humans expanded their number and reach, the influential human footprint grew bigger and broader. Agriculture now constitutes the greatest human impact on Earth, covering over 5 billion hectares of the 13 billion hectares of land available (see www.fao.org). That is 5 billion hectares of displaced or eliminated species, usually replaced with species and genotypes unknown to the Earth of 10 000 years ago.

The term biotechnology can also be applied to many traditional and modern forms of food processing: making bread by combining yeast with crushed wheat, wine from grape squeeze and microbes, beer from hops, barley and yeast. These forms of food production using fermentation technology require human intervention and manipulation of nature to yield the food products most humans have enjoyed for millennia. Almost all other foods and food processes have been altered, modified, or improved by humans over history (see, for example, McGee, 1984). Today, apart from a few wild berries, some aquatic species that manage to survive overfishing, and some game animals that manage to survive increasingly techno-accelerated hunters, almost none of our common foods can be said to be genetically unmodified by humans (McHughen, 2000).

More modern forms of plant breeding and animal husbandry have generated genetically modified forms of crop varieties and animal breeds. These novel beasts and plants carry traits suited to human cultivation and consumption, not to fitness and survival in the wilderness, as Nature would have. And it is true that human meddling in the genes of plants and animals, even using "natural" means of genetic modification, can and does lead to unintended effects, even mistakes requiring postmarket eradication (NAS, 2004).

In modern parlance, biotechnology generally refers to modern genetic technologies, particularly recombinant DNA (rDNA), also known as genetic engineering. To gain a basic understanding of genetic engineering, one must understand the four basic concepts that serve as the foundation of genetics:

1. All organisms are made of cells and cell products
2. Each cell in an organism contains the same set of genes
3. The genome contains all the genetic information necessary to make an entire organism
4. All organisms share the same genetic language.

These four concepts encompass our knowledge of molecular and cellular genetics dating from the early twentieth century to the 1960s. For genetic engineering, the key is concept number 4. The fact that all organisms share the same genetic language allows a gene from any one organism or species to be read and understood when transferred to any other.

By analogy, consider a gene to be a recipe, and the genome to be the comprehensive encyclopedia of recipes, comprising all of the genetic information in a given organism. There may be 20 000–30 000 genes in the genome of a plant or animal, including humans. Each gene conceptually codes for a particular protein. When a gene consisting of the coded recipe for, say, the protein insulin is copied from the human genome and transferred to bacteria (which ordinarily lack the insulin gene), the bacteria acquire the ability to synthesize human insulin.

How is biotechnology (rDNA) used?

One of the problems in discussing the risks and benefits of biotechnology is the diversity of procedures generally categorized as "biotechnology" and the uses of them. Even rDNA, with its fairly strict technical definition (of recombinant DNA used to transfer genetic material from one species to another) encompasses diverse methodologies, which are unlikely to carry identical risks. For example, plant genetic engineers employ at least two highly diverse mechanisms to transfer DNA into plants. One is a biological method, based on the bacterium *Agrobacterium tumefaciens* as a naturally occurring genetic engineering agent, the other is a purely physical method, biolistic or particle acceleration. *Agrobacterium* is a soil microbe with the natural ability to transfer portions of DNA from the bacterial cell into plants cells, and have the transferred DNA permanently integrate into the plant genome. (*Agrobacterium* also indisputably refutes the claim of some that nature "never transfers genes from one species to another.") In the lab, scientists provide *Agrobacterium* with the DNA carrying the gene of interest, then inoculate the desired plant with the *Agrobacterium*, and allow the bacteria to do what comes naturally. The scientists then simply select the plants successfully transformed by the *Agrobacterium*.

The biolistic method, in contrast, uses no such biological assistance to effect the transfer. Instead, the DNA carrying the gene of interest is coated onto microscopic shotgun pellets, then the DNA-coated pellets are blasted into the plant tissues. Some cells will integrate the DNA into their own genomes, and then grow into a whole plant containing the gene of interest.

The diversity of these methodologies (and there are others also) ensure that any risks will not be common to all. For example, if there is a risk that rDNA might inadvertently transfer undesirable genes in addition to the genes of interest, such a risk would presumably be greater with the *Agrobacterium* method, which naturally transfers portions of its own DNA, than with the biolistic method, where the gene of interest is the only DNA available for transfer.

Applications of biotechnology

Genetic engineering techniques were first developed in the early 1970s and rapidly adopted for various commercial purposes.

The first was the commercialization of an example given above, the transfer to a bacterium of a human gene for insulin, to produce human insulin. Today, the majority of insulin used by diabetics is genetically engineered and produced by bacteria, instead of extracting insulin from farm animals. Other applications followed quickly, including a wide range of pharmaceuticals, foods, and crops. The first commercial food application was chymosin, a genetically engineered enzyme used in cheesemaking and produced by bacteria as an alternative to rennet, which comes from animal sources. The development of genetically engineered chymosin allowed true vegetarian cheeses and also, like insulin, saved unnecessary slaughter of farm animals.

The first genetically engineered wholefood crop was the now defunct Flavr Savr tomato, developed by the small California biotech firm Calgene and promoted as a longer shelf-life tomato, to provide (in their words) "summer fresh taste" in January when fresh tomatoes were at a premium in northern winter groceries. Since this release in 1994, a series of other genetically engineered crops have been approved and released for commercial production. These include herbicide-tolerant varieties of corn, soy, canola, and cotton, disease-resistant papaya and squash, and insect-protected corn, soybean cotton, and canola. Some of the commercial genetically engineered varieties were highly successful, while others, like the Flavr Savr tomato, failed and are no longer grown.

The biggest success stories are the major US field crops. In 2005, genetically engineered soybeans accounted for 87% of the total US soybean acreage, genetically engineered cotton claimed 79% of the cotton acreage, and genetically engineered corn was grown on over half of the corn acreage. In addition, genetically engineered canola was grown on three-quarters of the Canadian and US acreage of that heart-healthy oilseed, and the virus-resistant genetically engineered papaya is credited with saving the Hawaiian papaya industry from the devastating ravages of the papaya ringspot virus. There can be no argument that the introduction of genetically engineered crop varieties has had a dramatic rise and impact on US agriculture.

Red and green biotechnology

Traditionally, "red biotechnology" is the term applied to medical processes, producing drugs such as insulin, dornase alpha, and

Betaseron, etc., whereas "green biotechnology" is applied to agricultural processes, producing herbicide-tolerant soybeans, Bt corn, and disease-resistant papaya, etc. But what about: vitamin C, beta-carotene-enhanced rice, India's "Protato," hepatitus B vaccine in banana, or reduced mycotoxin in Bt corn?

Why is the distinction between red and green biotechnology necessary? Ethical principles demand that condemning one use of a "forbidden" technology must condemn all uses. Yet those arguing that green biotechnology is ethically questionable rarely raise the same question about medical uses of the same technology. The situation here differentiates between using rDNA to increase crop or food production, and using rDNA to generate medical treatments, especially such treatments as dornase alpha, for which there is no non-rDNA alternative. To argue that it is acceptable to abandon one's principles to support an unethical practice when that practice provides medical benefits, but not when the same practice provides food, is spurious and shallow. Clearly, food is a life-saving product for many poor people around the world, as important for sustaining life as any medicine in developed countries.

In the past, distinguishing agricultural applications from medical applications of rDNA technologies has been relatively easy, as it is a simple matter to draw a line between, for example, "herbicide-tolerant soybeans" and "insulin." But recent innovations have blurred this simple dichotomy. Most vitamin C tablets are now synthesized from corn, and over half the US corn crop is of genetically engineered varieties. The fact that the chemical composition of the vitamin C is identical, whether coming from genetically engineered corn or traditional corn, is irrelevant, because we are dealing with ethical issues of the use of rDNA. The point here is that vitamin C cannot be readily assigned to the (ethically acceptable) medical or the (ethically unacceptable?) food category. Other products of rDNA applications are similarly obscured: foods with enhanced nutrient content clearly have both medical and food value. And foods modified to deliver medical agents, such as vaccines, cannot be readily categorized as exclusively medical or exclusively food. Finally, the medical benefits of Bt insect-protected biotech corn (due to the reduced incidence and content of mycotoxins) are apparent, yet the rDNA corn was developed for agronomic value.

These examples show that the easy dichotomy between the ethically sound "red" biotechnology and ethically unethical "green" biotechnology is not sustainable, and the ethical distinction cannot be logically supported.

Biotechnology has been compared to a train

In this analogy, the US, Canada, Argentina, and China are on board and driving the train forward (Figure 2.1). Others are on the platform deciding whether to jump on or not. Of course, no one is sure where the train is going; perhaps to a better place, or perhaps into a chasm because the bridge collapsed. Before jumping on, prospective passengers need to decide where they want to go. If they choose a simpler community, less technology intense, then this train is clearly not for them.

One fear component in public understanding and acceptance of biotechnology is clearly the feeling of being "railroaded," that consumers are feeling pressured to board the GMO train when they may be reluctant to go anywhere, and even if they are, they are not entirely sure where this biotech train is going.

Figure 2.1 The GMO train (photo still from *In the Heat of the Night*, Mirisch Corp, 1967)

Risks: real and perceived

One consequence of presenting society with a highly complex and erudite technology is that misunderstanding runs rampant, and this leads to common misinterpretations with frightening overtones. Benign misinterpretations also occur, but without the fear factor these quickly disappear. Common scare stories derived from misunderstanding the technology include the following:

1. Genetic engineering breaks the "species barrier"; "Nature never allows genes from one species in another."
2. Genetic engineering involves random insertions into the genome.
3. Genetic engineering crops and foods are untested and unregulated.
4. Once released, GMOs can never be recalled.
5. There are uncertain future "unintended consequences" and hazards.

Such fears generate considerable opposition to biotechnology, but in many cases are either flatly wrong (e.g. 1 and 3) or are factually correct, but taken so far out of context (i.e. relative to non-biotech status quo products) as to give an incorrect impression (e.g. 2, 4 and 5). Nevertheless, such claims are frequent and do have compelling influence on a wary public.

Nanotechnology, as an emerging, highly complex technology, is also susceptible to public misunderstanding. Indeed, many of the same fears emanating from biotechnology are now appearing in regard to nanotechnology. Box 2.1 illustrates this with some of the concerns related to nanotechnology (see, for example, Johnson, 2005) that could grow into fearful, but not necessarily realistic, scenarios.

A number of studies have helped to clarify the issues and provide a starting point for analyzing risks and hazards. Although these are derived from biotechnology issues, they also relate to nanotechnology.

The risks can be classified as scientific (concerned with the environment and health safety) or non-scientific (ethical/cultural, socio-economic, political, covert trade, covert technological). The questions associated with the scientifically founded risks are: How will the biotech product affect the environment, particularly ecosystems? and How will the biotech product affect human and animal consumers? As a result, the science-based regulatory scrutiny of biotech products focuses on these areas. In the US, the Department of Agriculture (USDA) is the primary regulatory agency charged

---

Box 2.1: Is nanotech on the same track?

- "What are the unintended consequences?"
- "Nanoparticles can build up in the brains."
- "What is frightening is how little is known about how the particles interact with the environment and human body."
- "(Nanoparticles) can act in completely new and different ways."
- "(Nanoparticles) might ferry toxins right past the body's normal defenses."
- "Horrendous social and environmental risks" (The ETC Group)
- "When GMOs meet atomically modified matter, life and living will never be the same." (The ETC Group)
- "Nanotechnology pose(s) health and environmental risks great enough to justify banning." (*Washington Post*, reporting on the publication of the Royal Society and Royal Academy of Engineering report on Nanoscience and Nanotechnologies in 2004, http://www.washingtonpost.com/wp-dyn/articles/A25675-2004Jul29.html)

with assessing environmental risks with biotech crops, and the Food and Drug Administration (FDA) is primarily responsible for evaluating risks to the food and feed supply. The Environmental Protection Agency (EPA) is another primary agency charged with regulating products of biotechnology; they are mainly concerned with the use of chemicals in the environment (McHughen, 2006).

The risks posed by the use of nanotechnology are similar, and will invoke similar concerns regarding threats to the environment and health. These technologies also pose similar risks that are not necessarily scientific or health related, but nevertheless are real, important, and must be addressed. These include threats to ethical and cultural norms, to economic well-being and to political standards, particularly as related to international trade and technological competitiveness.

One of the reasons biotechnology faced problems in public acceptance and adoption was the recent addition of public involvement in technology decision-making. Through history, when a scientific development was introduced to the marketplace, it either failed or succeeded based on public adoption. If the risks associated with the new product or technology were seen as reasonable relative to the derived benefit, the product or technology met success. During the twentieth century, however, the risk issues started being addressed prior to market release, to assure consumers that putatively hazardous products are actually safe for release. In general, consumers were satisfied that the public scientists and regulators were doing a reasonable job and conferred upon them a high degree of trust and credibility in their decisions. In recent years, however, that trust has been eroded as a result of a number of high-profile errors or mismanagements, exemplified by the so-called "mad cow disease" outbreak (BSE) in UK cattle, giving rise to the dreaded variant CJD in humans. As a result, many consumers now demand a higher degree of accountability from public scientists and regulatory agencies, and a greater public involvement in the risk evaluation process.

Risk evaluation can be broken down into three components (Table 2.2):—risk assessment—, the actual documentation and analysis of risks associated with a new product or process, which is typically the job of scientists competent in that field; —risk management—, the monitoring and mitigation of the risks, which is the job of regulators; and—risk communication—, which arose in recent years to satisfy a public wary about the functionality of the first two. While the risk assessment and risk management duties are clearly delineated, the job of risk communication is not. No one is charged with the job of explaining to the public the results and implications of the risk analysis and management relating to a new product or technology. At the same time, implicit in the responsibility of all public academic scientists and servants is the duty to communicate, directly or indirectly, to society at large.

One major mistake by the public biotechnology community was the failure to fulfill the obligation to explain to the public what

Table 2.2 Traditional approach to risk	
Component	**Responsibility**
Risk assessment	Scientists
Risk management	Regulators
Risk communication	Who participates: No one ... Everyone ???

biotechnology was (and was not), what were the true risks and what mitigations were put in place to minimize the realization of those risks. As a result, the public was much more susceptible to misinformation and fearmongering, as they had little foundational means to authoritatively counter it.

Nanotechnology can preclude much of this problem by specifically delegating responsibility for risk communication to credible public experts.

Distinguishing perspectives

Many scientists express frustrations in dealing with the non-technically trained public, even those academics sensitive to and willing to help inform the public. Similarly, many non-scientists sincerely seeking information are frustrated by the attitude of scientists, sometimes misinterpreting this as arrogance or condescension. Instead, scientists and non-scientists should be aware of the dichotomy in thinking (e.g. see McHughen, 2002). When it comes to risk assessment, this dichotomy reveals itself in several forms. Scientists tend to look and think along pragmatic, science-driven lines, while non-scientists tend to think along ethereal, values-driven lines. These differences manifest themselves in a number of ways, as summarized in Table 2.3.

The pragmatic scientist evaluating risk in a new product or process concentrates on actual risks, compares the new product or process with current similar product or process, applies objective tests, and may conclude that the new product or process is "as safe as ..." the status quo counterpart—the product or process currently used. In contrast, the values-driven analysis adds perceived risks to the real ones, tends to apply the subjectivity of the precautionary

Table 2.3 Modern approach to risk assessment

Science driven	Values driven
Real risk	Perceived risk
Substantial equivalence	Precautionary principle
Objective	Subjective
Relative—"Prove it as safe as ..."	Absolute—"Prove it safe"
Product	Process

principle (not to be confused with ordinary prudent precaution), does not consider and compare the risks inherent in the current status quo product or process, and desires an absolute conclusion that the product or process is "safe."

One simple criterion to distinguish these two approaches is to consider "product" as opposed to "process." Scientists assert that hazards are associated with products, and that the process used to generate the product is irrelevant. That is, there may be several different processes that generate the same or very similar product. All of those similar products will carry the same risks and hazard profile, even if the processes used to create the products are distinct. A concrete example from agricultural biotechnology is the use of genetic engineering to generate an ALS-insensitive (herbicide-tolerant) canola variety. The risks associated with this variety are that the herbicide-resistance gene might escape into weedy relatives and persist in the environment, or that the new herbicide resistance makes the canola toxic and unfit for dietary consumption. Meanwhile, the process of traditional plant breeding can and has been used to generate an ALS-insensitive (herbicide-tolerant) canola variety. Although the processes used to generate the new canola crop are vastly different (even to the point where some say one process is so risky it should be banned, but the other is absolutely safe and needs no risk assessment or regulation at all), the fact is that the risks and hazards of these two canola varieties are identical.

A common assumption in the public debate is that "traditional" plant breeding is, by its very nature, risk free (or at least so low in risk as to be negligible) and, in contrast, breeding using genetic engineering is inherently risky. To test this assumption, the US National Academy of Sciences/Institute of Medicine (NAS/IOM) struck a panel to investigate this as part of a wider study investigating the health effects of genetically engineered foods (NAS, 2004). The study compared a range of different methods of plant breeding, including several "traditional" methods and some genetic engineering methods, evaluating each for the potential for unintended effects, a measure of potential for risk. The study concluded (as shown in Figure 2.2), that there were indeed differences in potential for risk among different methods of plant breeding. However, the differences were not between "traditional" and "biotechnology" methods. Instead, the potential for risk varied, with the least likely being a simple selection from a uniform population of plants, to the greatest likelihood of risk being another "traditional" form of plant

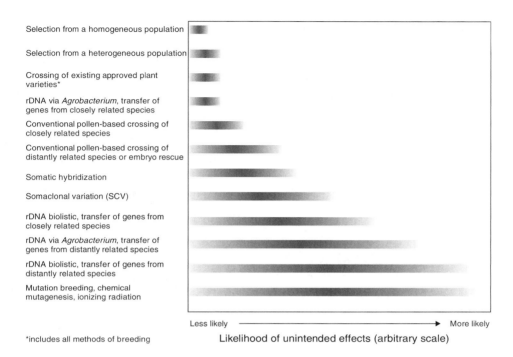

Selection from a homogeneous population

Selection from a heterogeneous population

Crossing of existing approved plant varieties*

rDNA via *Agrobacterium*, transfer of genes from closely related species

Conventional pollen-based crossing of closely related species

Conventional pollen-based crossing of distantly related species or embryo rescue

Somatic hybridization

Somaclonal variation (SCV)

rDNA biolistic, transfer of genes from closely related species

rDNA via *Agrobacterium*, transfer of genes from distantly related species

rDNA biolistic, transfer of genes from distantly related species

Mutation breeding, chemical mutagenesis, ionizing radiation

Less likely ⟶ More likely

*includes all methods of breeding

Likelihood of unintended effects (arbitrary scale)

Figure 2.2 Contrasting the objective, scientific approach to determining risk with the subjective, values-driven approach

breeding, induced mutation breeding. Several forms of genetic engineering breeding fell between these extremes, with one being among the least likely, and another among those with greater likelihoods of generating a risky crop. In all cases, though, the likelihood was extremely remote, showing that all forms of crop breeding, including biotechnology, are among the safest technological activities in modern society.

Nanotechnology as a process (or category of processes) may also be relatively safe, but the lesson to be learned from biotechnology is the context of risk. Like products of biotechnology, nanotechnology will supplement current technologies in providing new products. A major problem suffered in the biotechnology debate is that the relative context is often absent, particularly in the public perspective. That is, the public rarely considers the risks (or benefits) of biotechnology vis-à-vis conventional or traditional technologies of agriculture. Instead, the technology is often seen, and debated, in isolation, away from the context or reality. To illustrate this difficulty, consider the following issues in discussing another controversial agricultural component, pesticides.

To some consumers, pesticides conjure up toxic chemicals invariably causing environmental pollution and contaminating our food supply, posing risk both to health and environment. But the context is often missing. We often fail to consider the risks to health and environment in the absence of pesticides. Farmers use pesticides to control insects, weeds, and diseases. Without pesticides, invasive weeds (mostly invaders or introductions from overseas) could overrun natural environments, causing ecological devastation far greater than the judicious use of regulated chemical pesticides. And without pesticides, crops and foods could become much more hazardous due to infiltration and contamination from potential disease vectoring insects, diseases, and the toxic chemicals deposited in infected foods.

Among other points, context is lacking in those believing that pesticides, as synthetic chemicals, are invariably nasty and toxic, while naturally occurring pesticides are invariably benign and safe. In reality, naturally occurring pesticides are far more toxic than synthetic ones, and we consume far more natural pesticides in our food than synthetic ones (Ames and Gold, 1999). And some of these can be highly hazardous. Even so-called organic pesticides, used by the organic industry, can be true health hazards; rotenone and pyrethrin, naturally occurring pesticides approved for use on organic crops in the US, have been implicated in Parkinson's disease and as a possible human carcinogen, respectively (Ames et al., 1990). Similarly, context is lacking among those holding the belief that all synthetic pesticides are equally hazardous. In reality, synthetic pesticides display a wide range of toxicity, from essentially benign to highly toxic and hazardous. Finally, context is required to recognize the toxicologist's axiom: the dose makes the poison. That is, to some people, any amount of a pesticide (or GMO) is too much, even if the amount is minuscule and harmless. Indeed, some chemicals highly toxic in high concentrations are beneficial in low doses.

The problem of context can be summarized using the example of pesticides in agriculture. Popular misconceptions are that:

- "Natural" products are invariably safe
- Synthetic chemicals are invariably hazardous
- Toxicology doesn't matter: all chemicals are equally hazardous
- Amount doesn't matter: any amount is too much.

Table 2.4 Science versus non-science	
Non-scientific approach	**Scientific approach (NB: not all scientists)**
Starts with conclusion, searches for evidence to support it (cherry picking)	Collects and analyzes all available evidence before (perhaps) reaching conclusion
Discredits alternative views	Actively seeks alternative interpretations
Often lacks context	Is his/her own greatest critic Applies critical thinking skills

The lesson for nanotechnology is to ensure that public debate maintains a reasonable degree of context, as the biotechnology debate lacks it all too often.

Finally, and crucially, the nanotechnology community should apply the lesson of scientific credibility and the importance of scientific epistemology. In the biotechnology debate, many arguments claiming to be "scientific" are not, and society would be well served to be able to distinguish a scientifically sound argument from a non-scientific argument posing as one. Several features of scientifically sound approaches are readily identifiable, without the public necessarily learning scientific detail and technical minutia. In Table 2.4, the scientific and non-scientific approaches are compared.

Nanotechnology, like biotechnology, demands not only public support but public skepticism and critical thinking. In order to think critically, the public needs accurate information. When people vote without being properly informed, they risk voting against their principles.

The larger public debate surrounding biotechnology was and continues to be fed by, on one side misinformation, scare stories, and emotional manipulation and on the other side by misinformation and unrealistic or exaggerated promises of various benefits.

But biotechnology is not a black-and-white issue, its products are neither entirely safe and risk free, nor entirely hazardous without redeeming qualities. Furthermore, the process of biotechnology itself is not hazardous, according to scientific assessments dating back into the 1980s (NAS, 1983, 1987; OECD, 1986; OSTP, 1986). Instead, it is the products resulting from the process of biotechnology that carry risk, and those risks will vary product by product. Like every other product of technology, they have risks and benefits. And, again like every other technology, the key to broad social adoption, acceptance, and overall benefit is the appropriate management of risks while nurturing benefits. Undoubtedly, products of nanotechnology are similar, in that substantial benefits may accrue to the appropriate deployment of nanotech applications, as long as the risks are investigated and properly managed.

For the most part, the risks of biotechnology have been successfully managed, considering that genetically engineered products (from insulin to cheese to corn) have been on the market for over 25 years, and there is still not a documented case of harm anywhere, either to the environment or health. The scientific community recognized the potential of harm from inappropriate application of genetic technologies back in the 1970s, and various academic studies and regulatory agencies have continued to monitor developments and change regulatory oversight, resulting in this unusually "safe" status, especially considering the growth, extent and range of products of biotechnology worldwide. However, there is a divided partition between the views of the scientific and regulatory communities on one side and the views of the public on the other.

This gap, or disconnect, has formed largely because the public does not have access to what it has a clear right to: accurate information with which to make informed decisions. This does not imply that the public "will embrace biotechnology if only they had the right education" as some scientists aver, it means that reasonable people, using the same data, will tend to come to the same conclusions (which would help close the gap). Alternatively, the public might come to a different conclusion, but in that eventuality it would be based not on scientific evidence, but on other equally valid political, social, ethical, or other considerations. In this latter case, the division would remain, but it would be a legitimate and substantiated divergence, and the scientific community would reasonably be expected to accede to the public position.

To illustrate, the scientific community might reasonably assert that, for example, embryonic stem cell research carries great potential benefit and few technical risks; the public may look at the scientific and medical evidence, agree with the scientists but then demure, saying stem cell research is an affront to human dignity and therefore unethical. In this scenario, there is a division between the scientific community and the greater public, but it is based not on ignorance but on differing priorities, and is therefore valid. In this scenario, the scientific community has to recognize that merely because something can be done does not mean it should be done.

Conclusion

What, then, can nanotechnology learn from biotechnology?

The problem

In the current situation the often negative public opinion of agricultural biotechnology is colored due to lack of accurate information, exacerbated by unnecessary apparent secrecy and quietude. The "stay below the radar" strategy adopted by much of the biotech industry, and increasingly by the nanotech industry, is understandable but misguided. The "If we keep quiet, maybe they won't notice us" attitude is actually interpreted by consumers to mean "They're trying to hide something! It must be truly fearsome!" Which, of course, explodes the initial purpose of trying to remain inconspicuous.

Negative result

This lack of transparency and lack of expert information leads to a resulting lack of trust in the experts, whether in the scientific or regulatory community. Clearly, if you have expert knowledge of a potentially hazardous technology or product, and you refuse to share that knowledge with the larger society, then obviously you are not to be trusted because "people" and "safety" are not your priority. Trust is a crucial aspect in the success of any industry and, once lost, is very difficult to regain.

Building trust requires dedicated, strategic and intense education and outreach.

Action required

The easiest and proper way for the nanotechnology community to avoid the problems experienced in the deployment of biotechnology is to provide accurate information and encourage critical, informed analyses. This is largely a public communications effort, something that scientists historically have found difficult, in spite of their considerable ability to communicate with one another.

So this is another lesson and challenge for the nanotechnology industry. The scientists and educators with nanotechnology expertise must be encouraged to conduct outreach activities with the interested public and trained to deal with the public at large (their educational needs and expectations differ from those of students!). In the 1980s and 1990s, few molecular geneticists actively engaged in public education exercises. Those few academic scientists trained in biotechnology who did venture out into town hall or other public meetings often ended up either ineffectually debating the merits and demerits of biotechnology (and in the process alienated a previously neutral, open-minded audience who actually attended to learn something), or else retreating back to the safety of the ivory tower having been accused of unethical practices, arrogance, condescension, elitism, or "being bought off by industry." Scientists engaging in public education on controversial subjects do need a thick skin to protect against the kind of tactics rarely experienced previously, even in apparently vicious debates in academic settings. Such preparations are not taught in graduate school.

The downside of sponsoring public education programs is that some people, once properly and accurately informed, will still oppose deployment of the technology. This is a clear lesson from the agricultural biotechnology debate, where many biotechnology aficionados declared mass public ignorance of biotechnology a major problem and that the problem "would dissipate if only they knew what they were talking about." Well, mass ignorance is certainly a major problem in the agricultural biotechnology debate, but some people, when well informed, can still oppose deployment of agricultural biotechnology. In these cases, the rationale is not always due to scientifically unfounded fears, but due to other, non-scientific reasons, such as the propriety of control of the food supply. The nanotechnology industry can expect some percentage of a technically well-informed public to remain skeptical of nanotechnology for reasons similar to those of the well-informed opposition to biotechnology. Informed opposition is a legitimate position in a healthy democracy.

At the same time, scientific fact is not subject to the vagaries of democracy. No matter how many people believe the Earth is flat, or that gene transfer from one species to another is inherently hazardous, it does not change the shape of the Earth or the risk status of genetic engineering. Those who believe the Earth is flat will likely not be convinced by rhetoric, but may be convinced by experience and evidence. Similarly with the risks of genetic engineering, "convincible" skeptics need accurate information, evidence, and experience to reconsider their views. Nanotechnology is another erudite technology requiring public exploration, accurate information, and scientifically valid evidence. The wary public does not endow trust and acceptance based on pleadings ("Trust us, we're the experts"); they do afford trust based on transparency, honesty, and accurate evidence and information delivered in language they understand.

In summary, the lessons to be learnt from experience with biotechnology are:

- Transparency and honesty
- Outreach and public education
- Media and public policy training for scientists
- Building and retaining trust!

Following on the transparency, effective public education in technical and other matters, the nanotechnology will build and retain trust from most of the public. Even those well-informed opponents will adopt an attitude of respectful opposition, a condition for meaningful dialogue if the attitude is mutual and reciprocated.

References

Ames, B. N., Profet, M., and Gold, L. S. (1990). Dietary pesticides (99.99% all natural). *Proc Natl Acad Sci USA* **87**, 7777–7781.

Ames, B. N. and Gold, L. S. (1999). Pollution, pesticides and cancer misconceptions. In: Morris, J. and Bate, R., eds. *Fearing Food.* Butterworth Heinemann, pp. 18–39.

Hallman, W. K., Hebden, W. C., Cuite, C. L., Aquino, H. L., and Lang, J. T. (2004). *Americans and GM Food: Knowledge, Opinion and Interest in 2004.* Publication number RR-1104-007. Food Policy Institute, Cook College, Rutgers—The State University of New Jersey.

Johnson, C. Y. (2005). One million nanotubes could fit into this period. *Boston Globe*, March 13. Cited at http://www.masstech.org/mni/globe_2_15_05.htm

McGee, H. (1984). *On Food and Cooking: The Science and Lore of the Kitchen.* Scribner.

McHughen, A. (2000). *Pandora's Picnic Basket.* Oxford University Press.

McHughen, A. (2002). Scientists from Mars, consumers from Venus. *Nat Biotechnol* **20**, 549.

McHughen, A. (2006). *Plant Genetic Engineering and Regulation in the United States*. Publication 8179. University of California. Available online at http://anrcatalog.ucdavis.edu/pdf/8179.pdf

NAS, National Academy of Sciences (1983). *Risk Assessment in the Federal Government: Managing the Process*. National Academies Press.

NAS (1987). *Introduction of Recombinant DNA-engineered Organisms into the Environment, Key Issues*. White paper. National Academies Press, 24pp.

NAS (2004). *Safety of Genetically Engineered Foods*. National Academies Press.

OECD, Organization for Economic Co-operation and Development (1986). *Recombinant DNA Safety Considerations: Safety Considerations for Industrial, Agricultural and Environmental Applications of Organisms Derived by Recombinant DNA Techniques* (aka "Blue Book"). OECD.

OSTP, Office of Science and Technology Policy (1986). Coordinated framework for regulation of biotechnology. *US Federal Register* **51**, 23302–23393.

The Ethics of Agri-Food Biotechnology: How Can an Agricultural Technology be so Important?

3

Jeffrey Burkhardt

What Can Nanotechnology Learn from Biotechnology?
ISBN: 978-012-373990-2

Introduction

We are beginning to see "ethical" challenges to products developed using nanotechnology or containing nanotechnology materials. Since nanotechnology, like biotechnology generally and especially food and agricultural biotechnology (agbiotech), is a relatively new technology that has potential for radically changing industries and life generally, it may be useful to revisit the two-decade-old ethical debates concerning agbiotech. Certainly nanotechnology and agbiotech differ in important features. Besides newness, however, they do share at least one other commonality: somebody is "against them." Even a casual observer (or Internet surfer) of food and agricultural technologies cannot help but note that ethical objections to agbiotech—genetically modified organisms (GMOs) and GM foods—abound. Nearly every individual or group that has or has had ethical concerns about agbiotech has made those concerns known, in academic publications, public forums, governmental documents and hearings, and perhaps most significantly, on the Internet. This scenario is starting to emerge with respect to nanotechnology.

So, whither nanotechnology ethics? It is perhaps too soon to tell. But in this chapter I outline arguments regarding agbiotech/GM foods in each of the four areas that are likely to be relevant to the ethics of nanotechnology. These are environmental safety, food safety, consumer choice, and "structural" issues such as corporate power and governmental oversight. Nanotechnology may face challenges in each area, though perhaps in not exactly the same terms. Nevertheless, the agbiotech ethics debates provide food for thought for reflections on nanotechnology ethics.

When we think about the ethics of agbiotech (or any other technology) we usually mean applying ethical reasoning or ethical principles to the subject at hand. The question is, is a technology ethically right (acceptable) or not, and why? This, of course, is shorthand for asking whether the development and use of the technology is ethically acceptable according to basic ethical principles. Here, I want to suggest that three points drawn from a look at agbiotech ethics in the safety, choice, and structural arenas are relevant to present and future assessments of the ethics of nanotechnology: (1) Some so-called ethical arguments about biotechnology are really more about scientific fact or methodology than they are about ethical principles. This is generally true of arguments about the environmental–ethical acceptability of GMOs and about the safety of GM foods. (2) In the realm of consumer preferences, choices, and control over GMOs, there is usually agreement about facts, but the question of acceptability differs depending on which ethical principle(s) is (are) applied. (3) A final ethical appraisal of agbiotech may mean deciding whether it is really biotechnology (and specific consequences regarding the environment, food safety, or individual choices) that should be the focus of ethical concern, or whether it is something bigger. Although rarely if ever seen in print, it is a common belief among proponents of agbiotech, especially those in business and government (and university) enterprises with significant financial stakes in agbiotech's "success," that environmental, safety, and autonomy issues are "proxy issues" for a deeper and more wide-ranging critique of modern, industrial, corporate political-economic systems. Indeed, even if some GMOs turn out to be environmentally safe and safe for human consumption, the development and/or use of GMOs necessarily violates people's rights or spawns injustice, because people do not have a choice—not about specific food products but about the "food and agbiotech agenda." Accordingly, the "ethics of agbiotech" that should really occupy our attention is the "ethics" of the political and economic institutions responsible for developing, marketing, and regulating agbiotech in the first place.

Environmental, food, and consumer issues arise because of the ethics (or lack thereof) of the major actors who control the agbiotech agenda and ultimately the global food system. The question for nanotechnology ethics in this regard is whether its economic and political structure will end up just like that of agbiotech, challenging us to think not only about ethical principles, but about the economic world wherein technology is more than just a tool for producing things, but for producing power and profit as well.

Two main ethical principles have informed the agbiotech debates. These are the utilitarian principle of "maximum social welfare" and the principle of "respect for rights" or autonomy. Utilitarianism, simply, defines ethical acceptability in terms of consequences, specifically, positive net outcomes of actions. If an action (or development or use of a technology) makes more people happy, or satisfies more of peoples' preferences, then that action is ethically acceptable. If there are some, even many, "losers"—unhappy people, dissatisfied customers—that is an unfortunate but justifiable trade-off. Utilitarianism, classically stated as "the greatest good for the greatest number" takes a decidedly "scientific," and in some people's eyes realistic, view of the kinds of calculations needed to arrive at the net-happiness or net-satisfied preference ethical objective. For every action, there are always winners and losers; the best we can do, ethically, is to make sure the balance sheet for a given society comes up positive. Frequently, utilitarianism defines the balance sheet literally: economic gains and losses constitute the "goods" or "bads" that we calculate. Sometimes, however, other kinds of goods are included in the utilitarian social calculations.

Rights principles, in contrast, fix on absolutes. If an individual person or groups of people have rights to something, there is no legitimate way to ignore or override those rights. Rights are "trumps" against other socially desirable goods or goals (Dworkin, 1978). Accordingly, there can be no trade-off, no "balancing," no calculating net gains. Even though an action may achieve utilitarian gains, or serve some greater social good, if individuals have certain rights against that action, it is wrong. Rights principles are based on the idea of autonomy—individuals are the best judges of their own well-being—and nobody else's calculations can override autonomy.

Other ethical notions have reared their heads in the agbiotech arena: for example, religious ethical views, maintaining that actions (or technologies or their products) should not violate God-given rules, or God's creation/natural order, plan, etc. Or, some have argued that "virtues"—the idea that there are unique human excellences of character that we should aspire to and encourage—ought to serve as criteria for judging actions. I will briefly discuss these two latter, somewhat less well-traveled approaches to agbiotech only briefly below. Suffice it to say that utilitarian and rights approaches have served as the primary platforms from which agbiotech has been (and presumably nanotechnology will be) judged.

The environmental ethics of agbiotech

The most frequently raised issue with agricultural biotechnology has been the potential for GMOs to cause long-term disruption of natural ecological relationships, necessary for sustainable, productive agriculture, and the health of the planet itself. The concern is shared by proponents and opponents of agbiotech; the Earth must be treated in ways that ensures the welfare of present people and future generations. It would appear that both believers in the legitimacy of agbiotech (to the extent they have thought about it) and environmental critics hold if not espouse a utilitarian ethic in this regard. The question is what does the "greatest good for the greatest number" mean to each of these parties.

Some environmentalists object to agbiotech simply because they have deemed the products of agricultural biotechnology "unnatural." I will discuss "naturalness" in more detail below, but the claim can be understood to mean that the nature of the products, or the ways they have been produced are such that ecosystemic health, and by implication the health and welfare of present and future human beings (and perhaps animals), will be threatened. Many critics are unwilling to accept an argument that agbiotech products, despite their test-tube origins, might actually improve health and well-being by positively altering the environment. Rather, the complaints are that GMOs have had and will have (and maybe must have) specific negative consequences in the environment. For example, GM herbicide-resistant crops and their herbicide "partners" are indicted for killing microflora in the soil—to the detriment of long-term soil fertility and hence food security (see Lappé and Bailey, 1998). GM insect-repellent plants are been blamed for endangering beneficial insect species, thereby disrupting ecosystems in ways bound to negatively affect people (Rice, 1999). Critics point to potential (long-term) negative human welfare effects, whereas proponents stress positive welfare consequences: herbicide tolerance means fewer agricultural chemicals in use, ultimately reducing food prices; insect-repellent plants similarly reduce costs of chemical applications and hence decrease farmer costs.

The urgency of many environment-based criticisms stems from the number of GM products slated to be deliberately released into the environment. Although the rate of increase in patents for such products has slowed, the number remains high (Halweil, 2000). No environmental catastrophe has occurred because of GM products now on the market. Yet, it remains a contention of critics such as Greenpeace, the California-based Center for Ethics and Toxics, the Union of Concerned Scientists, and others that GMOs have the potential for causing serious ecological harm. These organizations and others have devoted considerable time and energy to showing how biotechnology products might behave in ecosystems, whether in a farmer's field or in the larger environment, with negative welfare implications. However, "mainstream" research and testing performed by governmental agencies responsible for monitoring and regulating products with potential environmental consequences have repeatedly stressed environmental safety (and agricultural/welfare benefits) (see, for example, Sears *et al.*, 2001). It is clear that much research is being conducted on environmental safety issues associated with GMOs, and different and contrary if not contradictory answers are being generated about their safety. This is controversial, but it is not an ethical controversy per se. Although, as some would have it, agbiotech puts the "ethics" of scientists and agencies who judge GMOs to be environmentally safe against the "ethics" of those whose scientific analyses suggest otherwise, the issue is really the soundness of the science used in establishing environmental acceptability (and preferability).

I see much of the environmental ethics of agbiotech as more of an apparent controversy than a real ethical one: it is more a case of disagreement over scientific interpretations and details, not ethical principle. Again, it would appear that researchers and activists working on assessing environmental impacts share an assumption or belief that "affecting natural systems negatively is ethically wrong," because of the welfare effects. Critics maintain that ecological diversity or soil fertility will be lost, negatively effecting present and future people. Proponents see improved gains in productivity and profitability (and reduced price). The difference is that the opposing sides have different views about what "the greater social good" means, and the means to achieve it. More important, they have different views about safety.

Proponents and critics of agbiotech seem to agree that all science can provide are its "best" conclusions at any point in time, and that there are gaps in scientific knowledge of how ecosystems behave, especially over the long run. Indeed, even the best scientific analysis of a new product is almost by definition incomplete. The question both sides pose is when are the results of scientific testing regarding environmental safety "good enough?" On the one hand, there are those in the scientific establishment, in biotechnology companies, and in government, who think that we can do the necessary tests and arrive at answers that are "good enough" to justify the release/marketing of a product. On the other hand, there are those who think that when doubts or uncertainties remain, we should continue testing until no more doubts remain.

This is the premise behind the call for scientists and regulators to adopt the so-called "precautionary approach" employed in environmental safety regulation in the European Union. Caution demands either near certainty in our environmental assessments, or at least serious consideration of "worst-case scenarios" (see Raffensburger and Tickner, 1999). US scientists and governmental agencies have not embraced the precautionary approach or precautionary principle. The standard view is that the kinds of rigorous chemical and biological testing of biotechnology products that "environmental risk assessment" mandates provide (and have provided) adequate grounds for asserting that some biotechnology products are safe.

A common definition of "safe" is "acceptable risk" (see Fischoff, 1981). Determining safety usually involves a two-stage process. First, scientific risk analysis is applied to a product. Risk analysis involves the identification of hazards or harms associated with the product, assessment of what effects a product has given different levels of consumption, and assessment of possible effects on different categories of agents, e.g. children, normally healthy adults, etc. (Wotecki, 1998). The goal of the process is a judgment concerning the "probability of harm"—how likely is it that this product will produce any negative health effects.

The second and crucial stage in the evaluation of environmental safety is deciding whether possible/probable harms are acceptable, and according to what standards. One such standard is "no detectable adverse effects." For example, if a chemical shows a low (though not zero) probability of harm over variable doses and different populations, and if there is some benefit in its use, applying the "no detectable adverse effects" standard means that the product is judged safe (i.e. the risks are acceptable). Alternatively, risk assessment may determine that for some populations, or at some dose, or in the presence of another substance, the probability of harm may be somewhat high. One might assume that risks would be unacceptable. However, another standard, "risk necessary to achieve benefit," might allow the product to be determined safe—though with conditions such as appropriate labeling, or applicator certification required.

Despite the apparent rigor in environmental risk assessment, opponents of GMOs fix on the fact that "safety" is a value judgment rather than a scientific certainty. As such, judgments of safety are to be treated with suspicion. This perspective leads some to believe that critics are anti-science, or worse, anti-rational and "emotional" (Rollin, 1995; Vanacht, 2000). However, the critics' points imply a deeper ethical indictment of risk assessment, and especially scientists' ethics, despite the latter's apparent assent to a "do no harm to the environment" principle. This concerns sincerity and honesty: How committed are scientists and regulators to environmental safety? How diligent are scientists and regulators in applying rigorous tests to determine safety? How careful are they to alert farmers to use agbiotech products appropriately? Each of these matters has bearing on the ethical acceptability of agbiotech, but are not typically the focus of the debate about the environmental ethics of agbiotech. They are relevant in terms of the institutional context of agbiotech's development and deployment, and I will return to this.

In sum, once we set aside those arguments that assert GMOs are environmentally unacceptable simply because they are GMOs, the environmental–ethical acceptability of GMOs rests on a judgment that these products are safe. This judgment rests, in turn, on someone having subjected particular GMOs to a battery of scientific tests. If we believe in the integrity of scientists and governmental agencies, and hold that environmental risk assessment as currently practiced is "good enough," then we ought to be able to conclude that GMOs overall are not ethically unacceptable. The issues are really whether our current standards for determining safety adequately assure our acting ethically responsibly toward the environment, and scientists and others responsible for safety determinations perform their tasks competently and ethically. Presumably, the goal of safety standards is welfare preservation and enhancement, a goal held by pro-agbiotech types and those opposed. Perhaps the issue, ultimately, is whether the testing/regulatory system in place (and those that populate it) performs in ways consistent with environmental safety and human welfare. The same can probably be said with regard to nanotechnology.

The safety of GM foods

Some observers of the issue of the safety of foods containing genetically modified ingredients (GM foods) also raise the question of "naturalness," and again I will postpone discussion of this. The more predominant concern is whether GM foods are safe, and most of the points made about ethics and views about science apply to the question of the ethics of GM foods. I think it is fair to say that nobody is against food safety, or alternatively, everyone thinks it is ethically right that the foods we eat should be safe. Although one might make the argument that people have a moral right to safe food, the principle here is similar to that about environmental safety, namely, a utilitarian–welfare criterion. A safe food supply contributes to the greatest social good.

As in environmental risk assessment, food safety determination also employs scientific risk analysis, followed by the application of a standard such as "no detectable adverse effects." For GM foods, one issue that has been raised concerns a criterion that the US Food and Drug Administration (FDA) applies at the outset of its assessment process. This is the principle of substantial equivalence: if, based on chemical and nutritional properties and appearance and taste, there is no obvious difference between a GM ingredient and its non-GM or "natural" counterpart, only minimal additional testing of the GM ingredient is necessary (i.e. no human tests are required). If a GM soybean and ordinary soybean are deemed substantially equivalent, then the GM soybean is nearly "automatically" judged safe. The principle of substantial equivalence has itself been the subject of considerable criticism and debate, because, critics argue, applying this standard may lead the FDA to miss important facts about the composition of some foods or ingredients. For example, GM herbicide-resistant soybeans have been approved for several years under the substantial equivalence doctrine. Critics allege that the soybeans tested were not soybeans that had actually been treated with the herbicide to which they are tolerant. It is, they argue, the application of the herbicide that renders the GM soybeans unsafe (PSRAST, 2000). Moreover, since substantial equivalence does not require human tests for allergic reactions, situations may arise in which, for example, a soybean containing foreign (e.g. brazil nut) genes may be deemed safe, even if those genes make the soybeans or soybean products hazardous for people with allergic sensitivity to brazil nuts (Nordlee *et al.*, 1996).

Similar to the environmental issues, part of the concern with GM food safety rests with the adequacy of science, or at least science as it is performed by those agencies entrusted to guarantee environmental and food safety. The call is again for more and better science, the adoption of the precautionary approach, or simply more ethical scientific practices. Some critics of GM foods have suggested that the FDA's use of substantial equivalence with respect to GM ingredients amounts to a conspiracy between the government and companies producing GM foods. I cannot comment on the truth of this claim except to say that if it were true, the ethical challenge would be a serious one. The main point here is that with respect to food safety, questions again have more to do with science and scientists than with basic ethical principles. No one is against safe food; everyone believes safe food is a social good. The ethical acceptability of GM foods turns on whether or not they really are safe.

To reiterate, both the food safety and environmental safety arenas are instructive in that they represent areas of concern in which there appears to be agreement on basic ethical principles—do no harm to the environment, do not risk human health—but disagreement on the best (practical) means to achieve those ethical goals. This does not mean that the arguments are less serious: indeed, many major ethical debates throughout history have turned on questions of what might be called policy as opposed to principle. It does mean, however, that the kinds of final judgments about ethical acceptability will differ in these two realms in significantly different ways from the kinds of judgments appropriate in the other arenas I now wish to address, namely, the areas of consumer preferences/choices, and the question of political–economic control over GMOs.

Ethics and choice

The issue of choice regarding GMOs and GM foods in particular rests with this question: Is it possible for people who do not wish to consume GM products to not consume them? That is, is it possible for them to avoid GM foods, and will it be so in the future? More to the point, is it ethically acceptable that people who may be opposed to genetic engineering or who prefer to refrain from consuming these foods may not have a choice in the matter? The answer to this question depends in part on factual, practical matters such as the long-term availability of non-GM foods. More important, it rests on the acceptance of one of the two competing and conflicting ethical principles, the utilitarian principle of "maximum social good" and the rights-based principle of "respect for the individual's autonomy."

Again, proponents of GMOs argue that not only is there nothing wrong about GM products, but that they contribute in important ways to the social good (NABC, 1994). If nothing else, GMOs provide for a more efficient and cost-effective food production system. For example, the "first generation" of agricultural GMOs have been touted for (1) increasing milk output without increasing dairy cattle feed consumption (e.g. bovine somatotropin, bST); (2) simplifying weed-control regimes through the single use of only one chemical (e.g. glyphosate tolerance); and (3) reducing the need for chemical insecticides altogether (e.g. Bt crops). These GM technologies increased production/productivity, generating increased farm revenues and reducing prices to consumers. If increased farm productivity and lower (or at least steady) consumer prices are good things (which proponents maintain they are), the agricultural GMOs are themselves ethically justified on that basis: they help achieve maximum social good as it is associated with food.

The "next generation" of GMOs are claimed to be even more in line with this ethical value or principle, in so far as there is expected to be even more direct benefits to consumers: better nutritional content, enhanced flavor, extended shelf-life, even "nutriceutical" foods—regular foods that contain medicinal properties (Beachy, 2000). These direct benefits, it is claimed, will provide even stronger ethical justification for GMOs.

The implications of the maximum social good justification for GMOs are straightforward. If scientific tests show these products to be safe, there is no legitimate reason for anyone not to use and consume GMOs, just as there is no legitimate, ethical reason, for example, for consumers to not consume tomatoes imported from Mexico (assuming safety). It is, in fact, in everyone's best interest to consume GM foods for the economic and (potential) health benefits they provide.

Given this orientation, the fact that some individuals do not want to consume GM foods suggests to some that they are ignorant or foolish or both, and that as such their preferences simply should not matter. In utilitarian terms, these preferences (certainly in the minority at this point in time) should be overridden. If genetically engineered corn products were the only corn products available on the market, and an individual preferred not to consume genetically engineered corn, then accordingly he or she simply should not eat corn. Eventually, it is assumed, the individual would see that there is nothing unhealthy or wrong with genetically engineered corn, and once again he or she would consume corn products. There is certainly no utilitarian/maximum social good rationale for catering to the wishes of an "ethically illiterate" and "scientifically illiterate" individual or minority. If GMOs are ethically acceptable, from this perspective that is all there is to it.

However, the principle of respect for individuals' rights and autonomy demands that this be viewed in nearly the opposite way. Autonomy means self-determination, and if our primary ethical responsibility is to respect self-determination, then individual preferences or choices cannot be written off or ignored for the sake of the maximum social good (Cole, 1998). In fact, according to the rights/autonomy principle, any "social good" has to be defined as providing or allowing individuals the freedom to choose. This freedom includes the freedom to avoid GM foods for whatever reasons the individual sees fit.

Some individuals who choose to avoid GM foods do so because they have concerns about science, for example, the adequacy of risk assessment. Others may echo the beliefs of some consumers of organic foods, rejecting GM foods for environmental rather than food safety reasons. Some people reject GM foods for other, perhaps deeper philosophical reasons. Ever since the EU bans on importing GM foods and food ingredients, analysts have sought to uncover the reasons for European resistance to GM foods. Food safety is certainly among their concerns, but other reasons that have been identified are (1) a generally cautious view toward new technologies (as noted, the precautionary principle originated in Europe); and (2) a cultural tendency to identify food as something special, even "sacred" (Thompson, 1997b). Put another way, there is a belief that using genetic engineering on foodstuffs somehow violates the naturalness, integrity, and wholesomeness of food. For cultures that place high ethical value on their wholesome rural lives, their cuisines, the naturalness and integrity of farm animals, or natural environments, the genetic (technological) modification of nature and foods is deeply unethical. It is little wonder that many people in the EU have been so cautious concerning the GM foods "revolution."

While it appears that most of the consumers in the US generally do not hold such beliefs, a similar point could be made about those who choose to consume only "whole foods"—fresh fruits, vegetables, grains, and meats—which are then prepared in the home. People may choose to do so because of the "naturalness" or freshness of meals prepared from fresh ingredients. Others may attend to the same sort of "sacred" aspect of meals carefully prepared for their children or families. In both cases, these consumers avoid preprocessed foods, and count on the availability of whole foods (even if availability is seasonal for some ingredients). The issue is whether these consumers will be able to continue to act on their values and beliefs, as agricultural and food biotechnology continues to develop. Already, analysts suggest that over half of soybeans and more than a third of corn planted in the US are GMOs. If this trend were to continue, they may be few if any non-GM crops grown in the US. If there are no non-GM whole foods available, and there are people who reject GM foods as unacceptable, unnatural or "fake," something is wrong. Not only are they put at a disadvantage, their rights/autonomy as consumers is violated. Is this fair? This is as (if not more) important than the case of a school cafeteria failing to offer vegetarian or kosher options to its (captive) patrons (Thompson, 1997a).

We are probably a good way away from the complete domination of foods by genetic modification. However, from an ethical point of view, the prospect of increased genetic modification of major whole foods suggests to adherents of the autonomy principle that violation of some people's ethical or religious values, or just simple preferences, may accompany the growth of agricultural and food biotechnology. One response to this may be that those values or preferences are old-fashioned, provincial, simplistic, even irrational, and that not that many people in the US hold those values anyway. This, however, ignores the basic fact: these people exist, need to eat, and, according to the autonomy principle, have a right to be respected for their beliefs as do any other people.

It is at this point that the "naturalness" issue merits brief discussion. As noted above, some people hold that GMOs are environmentally unsafe or unsafe for consumption because they are unnatural. There are two ways to interpret this assertion. One is a quasi-scientific claim; the other is based on a philosophical/religious belief. According to the first view, because GMOs are the result of human intervention (biochemical engineering) and not natural selection, they are not only fundamentally different from non-GMOs, but will inevitably behave in environments or in the human body in unforeseeable and probably damaging ways (NLP, 2006). The standard response to this claim is that humans have been intervening in nature for millennia: indeed, all domestic animals and plant cultivars are the result of intervention. Genetic engineering is no different in kind from traditional plant crossbreeding or hybridization. In fact, the precision with which specific genetic sequences can be transferred to a new organism implies that we know more about how GMOs will behave in the environment than we do about non-GMOs. The issue is not the "artificiality" of a GMO but instead whether we can be assured that GMOs are safe. This "unnaturalness" objection presumably can be answered through more and better scientific analysis.

The other sense of "unnatural," though, is more pertinent to the current discussion of consumer autonomy/choice. According to this view, GMOs are unnatural because they are contrary to the "natural order," the way things ought to be. Nowhere is this clearer than in the philosophical and religious objections to transgenetic engineering: moving genetic material across species boundaries. In this sense GMOs are unnatural not only because they could not occur without human intervention, but because they are contradictory to Nature or perhaps God's plan. The implication of this is that humans should not "play God": genetic engineering (biotechnology) is simply immoral (see Rollin, 1995).

There is no appropriate scientific response to this belief. The common reply that we have been playing god for millennia (through selective breeding), fails to address the essence of this position (see Hansen, 2000). According to adherents of this belief, there are no analogies in anything else we do or engineer. Violating the order of nature in this way is simply wrong.

The fact is that many people who object to GMOs for cultural or philosophical reasons hold this belief. Again, to assert that they are simplistic or even irrational misses the point. This is their philosophical position; respect for autonomy necessarily implies that their views must be respected (or, at least, opposite views should not be imposed on them by restricting their opportunities to avoid GMOs). This is why the issue of consumer choice concerning GMOs is in many respects a paradigmatic confrontation between the competing ethical principles of utilitarianism and respect for autonomy. The principle of maximum social good asserts that if a product does no harm, and produces some desirable social outcome, it is ethically acceptable (perhaps even mandatory). Respect for autonomy demands that individuals be allowed to choose what products they consume or are exposed to, even if that runs contrary to the apparent social good. Given this clear dichotomy between ethical positions, determining the ethical acceptability of GMO with respect to consumer choices/preferences seems to demand that we reflect on the implications of the conflicting ethical principles, and choose one. Once chosen, a final ethical judgment about GMOs follows straightforwardly from that principle.

Ethics and control

The underlying (and ultimately practical) issue regarding rights/autonomy is whether non-GM foods will be available in the future, and whether consumers will have the option of refusing GM foods and still maintain nutritionally complete diets. This leads to a final set of issue that affects the ethical acceptability of GMOs. This area of concern is essentially the matter of who controls the food system, which means in part, who controls the agenda regarding GM foods or agricultural biotechnology inputs? By controlling the agenda, I mean, who ultimately determines what kinds of food products are developed, approved and marketed, and more basically, whether or not non-GM foods will continue to be made available to consumers—and farmers the world over who might choose not to grow GM crops. What are the ethics of such control?

Well before the matter of consumer choice and rights with respect to GM foods became subject of intense debate, ethical issues associated with the effects of agbiotech on small, in some case traditional-style "family farms" were raised (Burkhardt, 1988). Small-scale family farms have been under economic and social pressure for decades (see Berry, 1977). The demands of an increasing urban, global, and "convenience-oriented" market increasingly has put smaller farm operations at a disadvantage. Economies of scale captured by larger, specialized, and highly industrialized agricultural operations—agribusinesses—and public policies encouraging high productivity and low consumer prices put smaller operations in a pinch. Much of the pressure on small farms has been due to what is referred to as the "technology treadmill": unable to achieve the levels of productivity associated with high-tech farms, smaller operators either "get big"—and adopt the newest, labor- and cost-saving technologies—or "get out," i.e. "fall off the treadmill." The concentration of the majority of agricultural production (and revenues) in the hands of fewer and fewer farm operations has been the result. Small farmers have resisted, but the economics of agriculture has been a powerful enemy.

For many such farmers, agbiotech was seen as just the latest manifestation of the pressure to get big or get out. Though touted by the agbiotech industry as "scale-neutral"—equally able to be employed by large and small farms—small-scale farms and farm organizations saw the potential that agbiotech would be the nail in the coffin: at more than one national farmers' meeting, owner-operators (even big-time operators) have been heard to lament that they are essentially in the employ of the biotechnology/chemical corporations that supply their inputs (personal observation, National Association of Conservation Districts meeting, 2002). When bovine growth hormone (bST), which was supposed to increase milk production without marginally increasing costs, was introduced in 1985, smaller scale dairy farmers were among its most adamant critics (Burkhardt, 1989).

Farmers' ethical challenges to agbiotech have run the course from arguments that there are moral obligations to "save the family farm" (Comstock, 1987) to claims that the concentration in agricultural ownership/control in a few agribusiness concerns threatens food security. Elements of rights-based ethics and utilitarian concerns run through these critiques, along with "virtue ethics"—the idea that family farms embody, espouse, and protect certain moral ideals (Berry, 1977; Burkhardt, 1988). What these ethical objections have in common is the idea that forces (read: business and government entities) beyond anyone's control have begun to hijack food and agriculture, and this itself is a serious ethical and political-economic matter.

Agricultural biotechnology and GM foods do not enter into the environment or the marketplace or the kitchen from nowhere and without reason. As noted above, agbiotech emerged because of a certain "logic." This is the logic of modern, high-technology production agriculture, which demands increasing productivity from farms to keep up with people's demands and the scarcity of resources that can be devoted to farming. To this can be added the "logic" of molecular genetics, whereby scientists are made capable of manipulating organisms at the level of DNA. Plants (and animals) and microorganisms beneficial to agriculture are capable of being engineered in laboratories, relatively quickly (compared to traditional breeding), and ultimately delivered to farmers. The other elements in this "logic" are money and power. Since 1980, genetically engineered organisms have been patentable. Indeed, the prospects of obtaining patents on GM crops drove an economic revolution in the agricultural inputs industry in the early 1980s: seed companies were either acquired by petrochemical and pharmaceutical corporations already in the biotechnology business, or were infused with venture capital to establish biotechnology research and development efforts in GM seeds (Figure 3.1) (Busch et al., 1991).

AgBiotechnology Patent Ownership – 2003

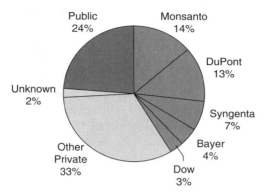

Figure 3.1 Agbiotech patent ownership, 2003. From Graff *et al.* (2003)

Global Maize Seed Market Share -- 2002

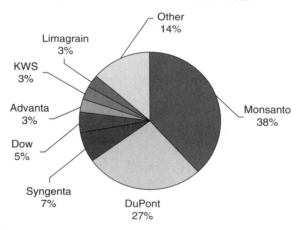

Figure 3.2 Global maize seed market share, 2002. From ETC Group (2007)

The process of acquisition, merger, and research redirection toward agricultural biotechnology has continued and accelerated. Presently, only a handful of multinational corporations control the global market for both GM seeds and agricultural chemicals (Figure 3.2). Economic concentration is the tendency toward monopoly or oligopoly in a given industry; it is safe to say that the agricultural inputs (seed, chemicals, machines) industry is sufficiently concentrated

(King, 2001; Graff *et al.*, 2003). The concern some people have is that agricultural inputs are not a "natural monopoly" in the sense that goods or services are most efficiently (and cheaply) delivered by single, large-scale enterprises such as electrical utilities. Rather, the concentration of the agricultural inputs industry (and indeed, the concentration in food production itself in the West) signals a tendency toward monopoly in what economists see as its most insidious form: ability to arbitrarily control prices and supplies. The fear is that this will be to the detriment of farmers, consumers and ultimately the public good.

Already, the power of patents, the concentration in market share, and the increasing unavailability of non-GM seed are causing global concerns. Consumer and farmer advocacy/activist groups have mounted transnational campaigns against what they see as gross injustices on the part of multinational (mostly US-based) corporations, Western governments, and the World Trade Organization (WTO). Patented GM seed, especially those containing "Genetic Use Restriction Technology" (GURT)—the so-called "terminator technology," they claim, undermines the sovereignty and viability of farmers in developing nations. Indeed, they argue that terminator technology was designed to control world agriculture. The ethical implications are clear: Terminator, "Traitor" (another GURT) and all such patented, multinational-controlled and US and WTO sanctioned agbiotech is ethically wrong (ETC Group, 2007).

The apparent ease with which multinational agricultural biotechnology corporations managed to secure patents and patent-like protections on GMOs and win approval from the US Department of Agriculture (USDA), Environmental Protection Agency (EPA), and FDA for marketing their GM products in the US, suggests that at least through the 1990s, corporations have indeed set the biotechnology agenda (Halweil, 2000). Indeed, despite the considerable degree of public uproar about such early agbiotech products such as bovine somatotropin (bST), and now Terminator seeds, these products have steadily made their way through the various national and global approval processes and are now on the market. Moreover, the fact that GURTs were jointly developed and patented by USDA scientists and the largest US seed company suggests the extent to which a US government agency, ostensibly charged to monitor and regulate agriculture and the industries associated with farming, have maintained less than an "arm's length" relationship with the biotechnology industry (RAFI, 1999). Added to these considerations, the United States' refusal to ratify international agreements limiting corporate power over biotechnology, or fighting them every time they arise in the WTO, implies further complicity between government agencies and corporate interests in promoting the development and use of GMOs worldwide. From the point of view of ethics, the question can be raised about whether such apparent corporate control over the biotechnology agenda serves the public good.

In 1999, increasing pressures on the FDA for the labeling of GM foods appeared as one attempt by consumer groups to assert control over the GM agenda, trying to ensure that it remained in the hands of the people. From an ethical perspective, labeling GM foods is consistent with both the principle of maximum social good and the principle of autonomy. Again, maximum social good dictates that GMOs should be available, and labeling does not hinder availability. It only gives people the opportunity to know what they are consuming, and allows them the option to choose freely. The doctrine of caveat emptor—let the buyer beware—makes no sense when there is no way consumers can determine the composition or ingredients of a product. GM foods are a case of such foods lacking "transparency" (Thompson, 1997b). Even if the ingredients of a product are "safe" according to FDA or USDA standards, the fact that consumers may want or need to avoid certain ingredients in a product (e.g. sugar, soy products) implies that it is ethically right to label the product accordingly.

The major objections to labeling GM foods were the costs of doing so, and the fact that doing so may stigmatize GM foods. Most analysts agreed that the cost factor would be small. So the real reason is that those who wanted to prevent labeling GM ingredients in foods is that they did not want any negatives associated with food or agbiotech. Opponents maintained that labeling would mislead consumers into thinking that GM foods are "different," when they have argued all along that GM foods are not (Biotechnology Knowledge Center, 1999). According to FDA regulations, labels are only required either when there is a risk factor (e.g. salt) or when a foodstuff or ingredient is fundamentally new and different. So GM labeling is not currently required.

The GM food labeling controversy in the US and in Europe was only the tip of an iceberg, however. Critics and governments across the globe have demanded that everything associated with agbiotech be labeled, if not banned altogether. Corporations involved in agbiotech have generally dragged their feet on such issues as "identity preservation"—ensuring that there is no cross-"contamination" of GM and non-GM seed. Some corporations have unilaterally withdrawn GM products in the face of global consumer and farmer complaints. By and large, however, the strategies have been to face ethical issues people raise regarding agbiotech with legal responses, e.g. challenges in the WTO, and lawsuits in the US and EU. The ways in which corporate actors behave vis-à-vis people's challenges or governments' regulations reinforces the notion that a major motivation on the part of the multinational "gene giants" is to control the global agenda regarding food. From whatever ethical perspective, the issue of corporate control over the food and agricultural system through control over its increasingly biotechnology-driven agriculture underlies and influences the other ethical concerns that have been identified. As important as environmental and food safety are, and as much as consumers may be worried about what goes into what they eat, if the "resolution" of those issues ends up as a decision at the headquarters of a multinational "gene giant," there is cause for ethical concern. There, the "greatest good of the greatest number" and "respect for individual rights" are seemingly irrelevant.

Conclusion: whither nanotechnology ethics?

I want to conclude this discussion with two questions: (1) How likely is nanotechnology to precipitate the kinds of ethical issues, and face the kinds of ethical challenges, that agbiotech has encountered; and (2) What ought to be the stance, or response, of the nanotechnology industry with respect to these ethical problems?

(1) I believe it is not only likely, but appropriate, that nanotechnology will face ethical critique, both public and academic. Technologies do not emerge or operate in social, cultural, or philosophical vacuums: goals, values, and intentions shape the technology, and the technology has impacts and implications. In the case of nanotechnology, we are already seeing issues raised concerning environmental and consumer (though not food) safety. A recent incident in the UK over a household cleaning product, touted as "using nanotechnology," that made several people ill illustrates that

safety considerations are simply going to arise. The ways in which nanotechnology products affect environments and impact human health will probably not be identical to those of agbiotech products. But there will be safety issues and concerns, and one can expect that the issues that will arise will be associated with the engineering "lineage" of the products ("naturalness"), the ways that ecosystems or human bodies are affected, and the nature and adequacy of attempts to keep those products safe "for the common good." As nanotechnology products become more consumer-oriented, in contrast to industrial, we should expect issues concerning choice and autonomy also to arise. We may not see the kinds of judgments that reject nanotechnology for what it is, in contrast to the "Frankenfoods" response GM foods evokes. Invariably, however, some people will prefer, and choose, to not purchase or use particular products of nanotechnology if for no other reason than that they don't understand the technology—or like being "forced" to adopt the "newest, new thing." Again, even if a particular product is really "good"—truly useful, economically efficient, environmentally benign, etc.—somebody is going to be "against it," and the ethical issue will be how to maintain freedom of choice with respect to it.

There is little doubt that structural, institutional matters will also surround nanotechnology—issues of power and control over the technology and the technology agenda. Like agbiotech, this may have little to do with the technology per se: the "logic" of modern, industrial, capitalistic political-economic systems is such that monopolization or at least concentration is the rule. Like agbiotech, nanotechnology may start and emerge from smaller R&D firms, or university laboratories, but to the extent that nanotechnology proves an economically viable and even transformative factor in our business and personal endeavors, multinational corporations will undoubtedly move in. In fact, corporate funding and in-house R&D already lead the way in the "nanotechnology revolution."

So, nanotechnology will face questions about its contribution to the greatest social good. Particular products will be challenged on the basis of potential environmental harm or threats to human health. Issues regarding individuals' rights to refuse and reject nanotechnology products will arise. And the nanotechnology industry as a whole will inevitably face an ethical critique of its structure, performance, power, and degree of control over both people's lives and the governments that are supposed to represent them. Agbiotech, unlike many technologies in the past, has had major problems with ethics that could not and cannot be ignored. I suspect nanotechnology will meet a similar fate.

(2) So what will be the response? The all-too-common response by the scientific community and by corporations involved in foods and agricultural biotechnology to challenges to GM foods is that the public needs "more information." The idea, evidenced by the high-power information and public relations campaigns being waged by corporations such as Monsanto and Du Pont, is that challenges to GM foods are the result of less-than-complete or accurate information about food biotechnology. The assumption is that potential widespread rejection of GM foods in the US, similar to what has occurred in the EU and the UK, can be avoided or mitigated by a science-based public relations effort. This attitude on the part of pro-GM forces has followed a pattern similar to responses to other occasions where there has been some public outcry or concern associated with new science or technology. Better "scientific literacy" is the key to general widespread acceptance of whatever the scientific community and the science (or life science) industry produces.

This attitude is not without merit. Time and again consumer attitude surveys have shown that acceptance of new technologies, faith in science (though mistrust in transnational corporations), and faith in governmental regulatory/oversight agencies, all increase as the level of education or quantity of information about a new technology increases (Hoban and Kendall, 1993; Saad, 1999). This suggests that strategies to gain consumer acceptance should be a matter of education or more and better information. The assumption seems to be that consumer acceptance (or lack thereof) and ethical acceptability (or unacceptability) are the same.

This is not the place to fully explore that distinction. Rather, I raise it because, in the small but growing body of public discussion of ethical issues associated with nanotechnology, it appears that the nanotechnology industry response parrots that of agbiotech: critics "don't know enough" about nanotechnology to offer reasoned critiques. This may be true. It does not, however, excuse the agbiotech industry from taking ethical challenges seriously, when they are principle-based and not simply "complaints." Nor should it, I would argue, excuse the burgeoning nanotechnology industry from confronting serious, reasoned ethical critiques and even anticipating potential ethical issues.

Whether or not an industry can take ethical challenges to its products and even itself seriously, especially one controlled by a very small number of large, multinational corporations, is a big question.

It has not happened often in agbiotech. Occasional conferences and workshops on "ethical issues in food and agricultural biotechnology" convened by corporate executives and researchers, or government regulatory agencies, suggest that it is not impossible or inconceivable to bring ethics "inside" the technology establishment. But these are exceptions. Whether they will be exceptions in the world of the nanotechnology business remains to be seen.

References

Beachy, R. (2000). An overview of GMOs. Paper presented at the Conservation Technology Information Center Forum, The Future of GMOs: Driven by Perception or Reality, Colorado Springs, CO, March 2, 2000.

Berry, W. (1977). *The Unsettling of America*. Sierra Club Books.

Biotechnology Knowledge Center (1999). "Agri-food Community Open Letter to President Clinton on Science-based Labeling of Foods," GENET news <genet-news@agroanet.be>, November 19, 1999.

Burkhardt, J. (1988). Biotechnology, ethics, and the structure of agriculture. *Agric Human Values* **5**, 53–60.

Busch, L., Lacy, W., Burkhardt, J., and Lacy, L. (1991). *Plants, Power, and Profit*. Blackwell.

Cole, P. (1998). *The Free, the Unfree and the Excluded: A Treatise on the Conditions of Liberty*. Aldershot.

Comstock, G. (1987). *Is There a Moral Obligation to Save the Family Farm?* Iowa State University Press.

Dworkin, R. (1978). *Taking Rights Seriously*. Duckworth.

Fischoff, B. (1981). *Acceptable Risk*. Cambridge University Press.

Graff, G., Cullen, S., Bradford, K., Zilberman, D., and Bennett, A. (2003). The public private structure of intellectual property ownership in agricultural biotechnology. *Nat Biotechnol* **21**, 989–995.

Halweil, B. (2000). Portrait of an industry in trouble. *Worldwatch News Brief* February 17. Worldwatch Institute. www.worldwatch.org

Hansen, M. (2000). Genetic engineering is not an extension of conventional plant breeding. Consumers Union Staff Paper. www.consumersunion.org/food/widecpi200.htm

Hoban, T. J. and Kendall, P. (1993). Public perceptions of benefits and risks of biotechnology. In: *Agricultural Biotechnology: A Public Conversation about Risk* (NABC Report 5). National Agricultural Biotechnology Council, Cornell University.

King, J. L. (2001). Concentration and monopoly in agricultural inputs industries. United States Department of Agriculture, *Agricultural Information Bulletin* 763. USDA.

Lappé, M. and Bailey, B. (1998). *Against the Grain: Biotechnology and the Corporate Takeover of your Food*. Common Courage Press.

NABC, National Agricultural Biotechnology Council (1994). *Agricultural Biotechnology and the Public Good* (NABC Report 6). Cornell University.

Nordlee, J. A., Taylor, S. L., and Townsend, J. A. (1996). Identification of a Brazil-nut allergen in transgenic soybeans. *N Engl J Med* **334**, 688–692.

PSRAST, Physicians and Scientists for Responsible Science and Technology (2000). Substantial equivalence versus scientific food safety assessment. Physicians and Scientists for the Responsible Application of Science and Technology Staff Paper. www.psrast.org/subeqow.htm

Raffensburger, C. and Tickner, J. (1999). *Protecting Public Health and the Environment: Implementing the Precautionary Principle*. Island Press.

RAFI, Rural Advancement Foundation International (1999). Genetic seed sterilization is the "Holy Grail" for ag. biotechnology firms. Rural Advancement Foundation International News Release, January 27, 1999. www.rafi.org/pr/release 26

Rice, M. (1999). Monarchs and Bt corn: questions and answers. *Integrated Crop Management* IC482-24. Iowa State University Cooperative Extension Bulletin, June 14.

Rollin, B. E. (1995). *The Frankenstein Syndrome: Ethical and Social Issues in the Genetic Engineering of Animals*. Cambridge University Press.

Sears, M. K., Hellmich, R. L., Stanley-Horn, D. E. *et al.* (2001). Impact of Bt corn pollen on monarch butterfly populations: A risk assessment. *Proc Natl Acad Sci USA* **98**, 11937–11942.

Thompson, P. B. (1997a). *Food Biotechnology in Ethical Perspective*. Blackie Academic & Professional.

Thompson, P. B. (1997b). *Food Biotechnology's Challenge to Cultural Integrity and Individual Consent*. Hastings Center Report 27.

Vanacht, M. (2000). Midwest growers survey regarding GMOs. Paper presented at the Conservation Technology Information Center Forum: *The Future of GMOs Driven by Perception or Reality*, Colorado Springs, CO, March 2, 2000.

Wotecki, C. (1998). Nutrition, food safety and risk assessment—a policy-maker's viewpoint. Paper presented at Purdue University, June 18, 1998. www.fsis.usda.gov/speeches/1998

Internet references

ETC Group (2005). Terminator seed battle begins: farmers face billions of dollars in potential costs. http://www.etcgroup.org/article.asp?-newsid = 548

ETC Group (2007). *World's Top 10 Seed Companies—2006*. Available from http://www.etcgroup.org/en/materials/publications.html?-pub_id = 615

NLP, Natural Law Party (2006). *Platform on Agriculture*. Available from http://www.natural-law.org/platform/agriculture.html

Saad, L. (1999). What biotech food issue? Americans not alarmed by application of biotechnology in food production. The Gallup Poll. Available from http://www.galluppoll.com/content/?ci = 3556

A View from the Advocacy Community

4

Margaret Mellon

Introduction

Like biotechnology, nanotechnology should not be considered inevitable. Although society may welcome the benefits of these two or other technologies, it faces choices in dealing with them. In particular, society has the right to regulate technologies—not only to protect against risks but also to enable those with broad and disparate interests to examine the products being developed and to consider alternatives.

The lessons that are to be learned from biotechnology depend on one's perspective. Nanotechnology, like biotechnology before it, promises enormous benefits for, but also potentially disturbing impacts on, society. And like biotechnology, nanotechnology could meet with resistance powerful enough to slow, if not derail, it. However, many of the obstacles that biotechnology encountered during the past 25 years proved desirable from a societal point of view, enabling many people who were once kept from that debate to participate.

What Can Nanotechnology Learn from Biotechnology?
ISBN: 978-012-373990-2

What can the proponents of nanotechnology learn that would make their technology more successful than biotechnology has been? How can critics of this technology do a better job explaining the risks associated with nanotechnology? How can they slow the technology enough to enable society to regulate and accommodate its impacts? How can we provide more people an opportunity to participate in the nanotechnology debate? How can society best modulate nanotechnology to reap its benefits and avoid its risks?

Basics of the biotechnology debate

Simply put, biotechnology consists of the practical use of living organisms. Much of the debate about this technology focuses on a subset of products created through artificial gene transfer techniques referred to as genetic engineering. The technology has broad potential applications because genetic engineering techniques can be used on virtually any living organism. Applications include research tools, drugs, crops, and food ingredients. The US government heavily promoted biotechnology in the 1980s as an important engine of economic development.

Concerns over biotechnology reflect the novel—and potentially irremediable—nature of the risks it poses. Those risks emanate from the unusual combinations of genetically determined traits—which are not practically possible in nature—that can be achieved matter-of-factly through genetic engineering. Concern about biotechnology is further intensified because modified organisms can reproduce in the environment and, once released, in most cases cannot be recalled.

Although the products of genetic engineering in general fall into familiar categories—drugs, crops, and enzymes—the technology has the potential for making an array of dramatically different products. In addition, genetic engineering raises philosophical questions about the importance of natural boundaries between species, the gradual conversion of organisms into machines, and our relationship with nature. These features give genetic engineering an intrinsic fascination for many, but are considered appalling by some of its fiercest critics.

Indeed, biotechnology continues to attract a great deal of attention and resistance, especially in the international arena. But the potentially serious risks and philosophical issues by themselves do not account for the fury of the global response to biotechnology. During the past 25 years, the debate has added dimensions. Besides concerns over risks, these dimensions include: (1) global resentment of American technical hegemony; (2) concerns about agricultural industrialization and corporate control of the food system, (3) blurring of boundaries because of increased university–corporate relations, and (4) lost opportunity costs as the biotechnology industry and allied scientists absorb resources, leaving other technologies under funded.

Continuing controversy for agricultural and food applications

How has genetic engineering fared in the 25 years since its debut? It has succeeded spectacularly in advancing biomedical research and has made major contributions to the pharmaceutical industry. But in the agricultural and food arenas, the product offerings are limited compared with the extravagant promises that accompanied its debut. Moreover, its momentum is slowing. Only two commercially successful traits are engineered into commodity crops: one that allows the use of certain herbicides and one that fends off insects. Crops containing these traits have been widely, although by no means universally, adopted.

Agricultural and food applications of biotechnology still face opposition and controversy. In many countries, resentment is entrenched. Even in the United States, where the technology is widely accepted, acceptance or commercial production of genetically engineered food animals appears to be sidelined. The next generation of genetically engineered plant crops involves minor changes or combinations of existing traits. There are very few new or established biotechnology applications that are directly attractive to food consumers, and therefore to food companies. On the contrary, many food companies are currently spending millions of dollars to avoid the use of such biotech-derived products.

The reasons for agricultural biotechnology's current state are complex. In part, the technology is genuinely new, emotionally disturbing to some people, and also presents potentially irremediable risks. Moreover, the industry blundered repeatedly when it tried to fend off criticism, often adopting an arrogant and elitist tone.

Further, because the technology sits at the nexus of so many issues, it attracted an exceptionally diverse array of opponents. And the technology proved difficult to apply to plants and animals, while it offers relatively few benefits to offset concerns about risks.

Classifying nanotechnology risks

Will nanotechnology be perceived as similarly controversial? Nanotechnology is defined by the scale at which molecules can be manipulated, and there appear to be two clusters of nanotechnology applications. One involves industrial products similar to standard chemicals; this includes applications such as coatings, sunscreens, and diagnostic probes. In terms of their benefits, these products represent incremental advances in existing technologies. Some of these products are on the market or will be soon. The second, more futuristic cluster is based on the convergence of biotechnology, nanotechnology, information, cognitive, and other advanced technologies. Those applications are considered revolutionary and could challenge what it is to be human. They also are described as offering fundamentally new platforms for the manufacture and distribution of goods.

The industrial applications of nanotechnology belonging to that first cluster appear unlikely to engender special societal concerns. In general, the mere notion of very small scale does not have overtones akin to genetic engineering and therefore is unlikely to be philosophically controversial. The technology may lead to applications or capabilities that demand a social or political response, but such impacts will be difficult to tie directly to size. In general, industrial nanochemicals do not sit at the intersection of as many social issues as does biotechnology. For example, they do not seem to provoke anxieties about weighty matters such as American hegemony, the nature of life, and control the food system. These novel products present a panoply of benefits and risks but, in general, they fit within the usual trajectory of progress. Moreover, nanoscale products in drug delivery, bio-imaging, coding, and computer chips do not seem readily distinguishable from comparable microscale products. These applications are exciting for someone in those economic sectors, but they seem incremental and unlikely to engender a special response from society—unless nanoscale manipulation leads to a novel class of health risks.

Of course, industrial nanochemicals could present health and environmental risks. They may be toxic, move into and persist in the environment, and affect organisms and ecosystems in unpredictable ways. Such risks are serious and not completely understood, but at this point are not the basis for carving nanochemicals out for special regulatory attention. Current chemical regulatory systems are deficient and should be strengthened and improved, but not simply to address the issues raised by nanochemicals.

Consequences if nanochemicals present special risks

The kinds of risks that necessitate special treatment are the unique toxicities that some experts believe nanochemicals possess. The suggestion is that molecules at nanoscale have different properties: they may penetrate tissues that other chemicals cannot reach, persist in the environment, or be more reactive than other chemicals. Despite a few provocative studies, the scientific community has reached no consensus on whether nanochemicals are fundamentally different from ordinary chemicals in terms of their risks to humans or to the environment.

Society should place high priority on resolving this question by funding substantial research programs addressed to this issue. If nanochemicals are uniquely dangerous, they should be treated specially and, if appropriate, even banned. If nanochemicals do not pose unique toxicities, they should go ahead, subject to the regulatory constraints on all chemicals. In the interim, the precautionary approach is to delay commercialization of new nanochemicals.

Some proponents of nanotechnology envision it bringing us nanobots, machine–human hybrids, or enhanced humans who are reliably brilliant, long-lived, and healthy. Such applications raise many of the same concerns about the meaning of being human and our relationship to nature as does genetic engineering, including the specter of reproducible and uncontrollable entities. But these visions are in the concept stage and are conceivable only if nanotechnology is combined with other technologies.

Since these two clusters of applications are so different in their time-scale, intrinsic interest, and transformative impact, the question arises whether the convergent and the industrial applications of nanotechnology are properly treated as one entity. It seems better to treat them separately. The answer has implications for how society will respond to the technology as well the best way to regulate and control it.

Questions about convergent technologies need to be addressed in a new way. The major issue is not a scientifically resolvable question about toxicity, but a fundamental question of where we want technology to take us. We should not leave such questions for the future but need to start talking now about what it means to be human, about whether society can choose a future, or will simply slide into one on the path dictated by technology. These questions do not make sense being framed in terms of industrial nanotechnology because technically oriented conversations about cosmetics and paint coatings obscure these more profound issues.

From experience in the biotechnology debate, it will likely prove difficult to separate these discussions because too many people benefit by entangling them. The futuristic issues are part of the glamour of the technology that is being used to establish it. That glamour is needed to elevate the profile of nanotechnology and to justify millions of dollars of investments. This part of the nanotechnology issue echoes the biotechnology debate. Both were technologies designated as engines of development and driven into the public consciousness with exaggerated versions of their economic and social potential. That hype is going to make it impossible to separate the issues in a way that makes sense.

Three lessons

For society, be wary of hype. Like nanotechnology, biotechnology debuted and persists in a cloud of hype, with promises of self fertilizing corn, chemicals-free agriculture, the abolition of hunger, massive increases in yields, food that is medicine, and other huge promises. What has been accomplished, although substantial, does not come close to that vision. For some, that is to be expected— entrepreneurs always oversell. But hype is a danger when it leads us to invest limited resources in the wrong place, which is exactly what happened for our agriculture-focused universities. For example, resources were diverted from traditional breeding, a far more powerful and successful technology for crop improvement than is genetic engineering. More worrying, society remains distracted from the intractable challenge of world hunger by embracing a technology with only marginal ability to address the problem.

The hype for nanotechnology, too, will alter priorities by redirecting funding. Those hype-driven priorities have consequences.

For industry, one lesson from biotechnology is to embrace strong regulations. The strength of a regulatory system is largely in the hands of industry. Government will rarely move regulatory systems beyond what industry will accept. In biotechnology, the industry wanted and got weak regulatory systems for most of its products. The system has saved the industry money and time in the short term but has cost the industry over the long term. For instance the US industry is hobbled in its international efforts to promote biotechnology because of weaknesses in the allegedly strong US regulatory system. Long-delayed recognition that the US system is largely voluntary does not engender trust among far-flung countries that are uncomfortable with US and multinational companies. In addition, the failure to put in place a comprehensive regulatory system left major applications of biotechnology, especially those involving animals, without a regulations-sanctified path to the marketplace.

Industry should consider strong regulations because they protect against risks, might reduce liability, and give society a chance to participate in key decisions. Participation can foster trust among ordinary citizens that proves necessary for moving controversial technologies forward. Because few people understand the technical details of biotechnology-related risks, they rely on government for information and judgment. The risks of nanotechnology will be no easier to understand than those of biotechnology. Strong, transparent regulations allow the kind of participation that engenders trust, while weak regulations that appear slanted towards industry contribute to feelings of unease, powerlessness, and cynicism.

For non-government organizations (NGOs) and activists, a lesson from biotechnology is to look for new institutions and approaches for bringing many voices into the nanotechnology debate, including consumers, organic farmers, ecologists, and many others. NGOs succeeded in bringing these and other voices to that debate, often through the media, consumer actions, international treaties, and the established regulatory system. But all those approaches have limits, and there are not enough institutions within which society can grapple with critical decisions about new technologies. The need for such mechanisms and institutions is a major challenge for all, but especially the NGO community, which is so often frustrated by the constraints of the regulatory system. Those institutions must offer genuine participation and must have the power to say no if they are to be considered legitimate.

In summary, the nanotechnology debate echoes the biotechnology debate in many ways. Both technologies encompass broad sets of products defined by process—in the case of biotechnology, the artificial transfer of genetic traits and, for nanotechnology, the scale on which molecules are manipulated. Both share the status of being government-created engines of development. Both present uncertain, potentially catastrophic risks, and raise concerns about societal control and consumer choice. The technologies differ in the intrinsic interest that they elicit, with nanotechnology dependent on the extent to which it appears to be connected to futuristic applications of advanced technologies.

Whether nanotechnology attracts opposition sufficient to impede its course depends in some measure on its proponents. The proponents of industrial nanochemicals have the option of separating themselves from the visionary applications, boldly resolving the issue of toxicity, moving ahead if the science allows. The first step down that path might be to walk away from the term nanotechnology and again embrace a prosaic name, something like modern materials science.

Questioning the Analogy (From Bio to Nano)

The Three Teachings of Biotechnology

5

Mickey Gjerris

Introduction

There are many suggestions these days that nanotechnology should learn how to deal with questions regarding society and ethics from the biotechnology experience (Roco and Bainbridge, 2002; Royal Society and Royal Academy of Engineering, 2004; European Commission, 2005). But asking the question "What can nanotechnology learn from biotechnology?" is rather like asking the question "What do you want?" If the question is not to be answered by a string of questions, such as "For now or in the long run?," "For dinner or out of life?" and "In general or just from you?," it has to be placed in a context. It is by no means clear what "nanotechnology" or "biotechnology" are. Further, the question "What can nanotechnology learn from biotechnology?" does not explicate what the purpose of the learning experience should be.

What Can Nanotechnology Learn from Biotechnology?
ISBN: 978-012-373990-2

It is not that there are no answers to these questions. But they are seldom explicated. They form a series of unmentioned, undiscussed, and even unrealized assumptions about the desired outcome of nanotechnological development. Very often the analogy between biotechnology and nanotechnology is drawn because it is believed that by looking at biotechnology, nanotechnology might avoid the "troubles"[1] that the huge attention to ethical concerns in connection with biotechnology has presumably led to (Sandler and Kay, 2006). Nanotechnology should learn how to avoid the ethical controversy that large sectors of biotechnology has been involved in. In other words: the question "What can nanotechnology learn from biotechnology?" can most often be translated into "How can nanotechnology avoid the ethical scrutiny that biotechnology suffers from?"

As so many are already attempting to answer that question, I have chosen to answer a different one within the same general debate: "What can we as citizens, as members of societies (stakeholders, individuals, citizens, researchers, etc.), learn from the biotechnology experience about ethically scrutinizing new technologies in the best possible way?"

This essay will focus on three distinct teachings extracted from my experiences of the discussion surrounding biotechnology. These teachings, I believe, can be used to illuminate the incipient debate about the ethical and societal implications of nanotechnology. These teachings are

1. To be able to have a meaningful discussion, it is crucial to know at what level of abstraction the problems at hand are being discussed—by oneself and by others.
2. Do not underestimate people who do not agree with you—rarely are they just uninformed.
3. If you want a dialogue then be aware what it is—do not continue the monologue.

The methodology applied in this essay is inspired by the continental tradition of phenomenology—especially the Danish thinker K.E. Løgstrup and his attempt to create philosophy as an ongoing interpretation of his immediate experience of the many phenomena in nature and life that are independent of humans, although we to a great extent depend on them—such as trust, beauty, love, mercy, life, and language.[2] It is philosophy at its most subjective, but perhaps also most useful form: thoughts on contemporary problems from an individual. The value of such thoughts lies not in their objective truth, but in the extent to which they are recognizable to others. Thus the methodology has been called *demonstratio ad oculos* (go and see for yourself if the world is not as I have just told you) (Jensen, 2001). This essay should be read as an invitation to walk out in the landscape of ethics and technology and see if there is something that can be recognized.

What are we talking about?

When ethical concerns about new technologies are discussed there is always a distinct danger that people will talk past each other. Not necessarily out of ill will or because one of the participants in the dialogue is lacking in knowledge,[3] but simply because they are discussing the matter at different levels of abstraction. To clarify this let us consider an example from the biotechnology debate (Peters, 1996) that is also cited sometimes as an example of the kind of concern that nanotechnology might have to face (Buerger, 2006), the concern that scientists might be playing God.

In the debate about biotechnology the phrase "playing God" often characterizes the ethical concerns about technology that cannot readily be reduced to concerns about negative effects on human health and the environment. The phrase indicates that the concerns to be discussed under this heading are metaphysical or religious in nature and not prone to be "solved" by the same technologically oriented rationality and framework that the more physical concerns are usually discussed within. This means that concerns of this kind are often listed last under the heading of "other moral concerns" and only briefly discussed (Gjerris, 2006). In this context "to play God" means to do things that humans are not supposed to—usually with humans or other biological entities through the means of new technologies.

"Playing God" is essentially a metaphor that points to concerns about the amount of power that new technologies provides humans with, concerns about the way this is changing the way we understand ourselves and the rest of the world and concerns about transgressions of some kind of natural order and inherent integrity in non-human entities. If the term "playing God" is understood this way it becomes clear that it is a culturally conditioned expression for a range of metaphysical concerns about humankind's relationship with both the inner and outer nature. These concerns are not readily discussed within a risk-oriented, technologically solution-based rational framework; they have to be discussed within a philosophical/metaphysical/religious framework.

Nonetheless it very often happens that the metaphor is taken literally and refuted as being contradictory and based on false assumptions. This is obviously not very hard as the notion of "playing God" taken literally can just be meant to imply it is somehow wrong that humans try to take their fate into their own hands. It can then be shown that it is in no way clear why this should be wrong or why this should make us especially cautious around new technologies. If changing the genetic composition of a plant is "playing God" (and thereby wrong), it needs to be stated why it is not "playing God" (and thereby right) to apply a band-aid or whether it is also ethically problematic to breed plants in the conventional manner (Dawkins, 1998; Kunic, 2003).

What happens here is clearly that different levels of concerns are being mixed up. On the one hand the proponents of the technology seek to show that what is done with the technology is just an extension of technologies already used, whereas the opponents seek to voice their concerns about some of the less tangible aspects of the technology. To argue that we have exercised power over nature for 10 000 years makes no difference to concerns about the novel powers that biotechnology gives humans. Something entirely different is at stake. As the American philosopher David Cooper (1998) has put it:

> Intensive farming of crocodiles is a new venture, but it represents an old wrong, now done to one more creature. By contrast, whoever first trained caged wild animals to perform acts at which audiences would laugh invented a novel kind of wrong, an addition to the already established list. Even in this case, of course, there were precedents, and if our criteria for novelty are sufficiently strict then, indeed, "there is no new thing under the sun". But then continuity with the past—similarities with what has gone before—is not the crucial consideration. The straw that broke the camel's back was just like the previous one in the bale, yet from the camel's point of view it was a very special straw. Sometimes, indeed, we only appreciate something as distinctive and novel by seeing it as the culminating stage—one that reaches a limit—of a continuous process. So the fact that genetic engineering of animals may be continuous with previous practices, such as dog-breeding or force-feeding, does not mean that it is innocent of committing a new wrong.

Discussing on different levels is not only time consuming but also futile. To the proponents of the technology, it seems as if the opponents do not grasp very basic facts about how humans live and act in the world and how we are already deeply embedded in technological manipulation of it, whereas the opponents are left with the experience of not being heard or taken seriously in their attempt to point to the perceived problems of the technology. This is not the best way to create a societal basis for a profound discussion of the way we as citizens would like technology to influence our lives.

One result of this discussion on different levels is the "knowledge deficit"[4] thesis that I will get back to in the next section, where, on one hand, proponents argue that opponents are largely missing the right information to accept the technology; on the other hand, opponents evidence a growing skepticism towards experts and authorities that are seen not as much as providers of information more as stakeholders with very definite motives and interests in the outcome of the discussion. A further result of this discussion on different levels is a risk that the whole notion of involving citizens in discussions and decisions about the implementation of new technologies will be met by indifference and cynicism by the very same citizens that one wishes to engage in the discussion.

So what can "nanotechnology" learn from this? I think the basic teaching to be had here is to listen closely to the concerns and accept that some of them might not be expressed within the risk analysis framework favored by both scientists and governmental agencies these days. If concerns are raised about the way nanotechnology will possibly lead to the production of some kind of self-replicating organisms or some kind of artificial intelligence, it is not the right strategy to dismiss these concerns as being unrealistic and give a longer lecture about the scientific limitations and reasons why this will never be realized. Rather, it is an opportunity to try to discuss the visions and motives behind the current development in an attempt to anticipate future developments and to influence the present (Grinbaum and Dupuy, 2004). The point is to understand these concerns not as literal worries that humans might turn into transcendent entities, but as metaphors that point to deep-held concerns about the direction our society is moving and the technological development that goes on. These are concerns that relate to a growing unease about the technological development and the increased power that humans gain over their own nature and the rest of nature as well. "Grey goo" might not be a practical risk, but understood as a flood of technology that will perhaps change the world, it can, as a metaphor, enter the important discussion of how nanotechnology should be developed.

The German philosopher Alfred Nordmann (2004) suggests that the nano-discussion should be seen as consisting of three different levels.

1. At the first level discussions concern the concrete risks and dangers that are raised by specific applications of nanotechnology (e.g. the toxicity of a certain nanoparticle).
2. At the next level discussions concern how and if the technology should be regulated to lead to the best results.
3. Finally, at the third level of discussions, "nanotechnology" is seen as a metaphor that carries with it all our hopes, expectations, and concerns about the future.

Thus nanotechnology can be said to consist of at least two different layers[5] that exist simultaneously but nevertheless very far apart.

W. S. Bainbridge (2002) has eloquently captured the double-headedness of the nanotechnology concept.

> There are really two different Nanotechnology Movements in the world today. One—based in industrial corporations, university laboratories, and government research-funding agencies—remains closely tied to chemistry, physics, and material science, working to create the actual technological breakthroughs of the coming decade or two. The other—based largely in science fiction literatures—postulates a future century in which nanotechnology revolutionizes human capabilities, based more on metaphor than on careful calculation, but having a profound influence on perspectives of people who are not scientists or engineers.[6]

The first teaching to be gained from biotechnology is thus: Be aware of the level of abstraction that the discussion of the technology is situated within—both regarding others and yourself.

If you do not agree with me you must be stupid!

Perhaps the most used litany in the whole of the biotechnology debate is that all that is needed to solve the controversies is more information and dialogue with the public. The interesting thing is that the idea that the disagreement about the development of the technology can be reduced to a question of lacking knowledge can be found on both sides: thus both proponents and opponents of the technology believe that more information will make their views stand stronger. As Reiss and Straughan (2000: 236) phrase it:

> Quite often those who advocate increasing public awareness and understanding seem to have a preconceived assumption that more "education" in this area will (or should) inevitably lead the recipients of this "education" to a particular set of conclusions.... For the "conspiracy theorists", for example, the end-result will involve the public coming to see through the machinations of politicians and big business and consequently to oppose or reject the new technology. For the "optimists" more "education" will (or should) lead people to see the benefits of genetic engineering and so accept it.

Most strongly the call for more information can be heard from biotechnology scientists who continue to believe that all they need to do is to get the truth through to the public that has been misinformed by the media. One noteworthy example of this can be found in a report called: "Why clone farm animals? Goals, motives, assumptions, values and concerns among European scientists working with cloning of farm animals" (Meyer, 2005). Here a group of scientists working with cloning is interviewed about the technology and various aspects of it. All of them say that it is very important to establish a dialogue with the public about the technology, but when directly asked what subjects would be suited for such a dialogue, they inevitably answer that the public should be provided with information about the technology.

Neverthless, there is no certain connection between a lack of knowledge about the technology and the acceptance of it. The knowledge deficit model is the idea that the pronounced public skepticism about certain types of biotechnology reflects a low level of understanding that can be remedied by feeding information into the public sphere more effectively. This model is challenged by surveys such as the Eurobarometer. The Eurobarometer surveys include a "knowledge quiz" in which respondents are asked a series of factual questions about biotechnology. This quiz enables the relationship between optimism about biotechnology and level of knowledge to be examined. The results show that when people acquire more information, they are better able to form an opinion for or against biotechnology—that is, there is a decrease in the number of "Do not know" answers. However, pace the deficit model, they do not acquire a more positive attitude to biotechnology. In particular, there is at best a poor correlation between knowledge and support of individual applications (Lassen, 2005; Lassen *et al.*, 2005). It seems more likely that the disagreements about biotechnology are fueled by different interpretations of the facts—interpretations that in the end reflect our values and visions of life.

Thus, the second teaching to take home from biotechnology is that it is possible to have an actual disagreement that is driven by values and not missing facts. What we should discuss is thus not what our opponents lack but the merits of our own visions.

A one-sided dialogue

The last teaching that I will touch upon in this essay is that it is important to bear in mind that there is a crucial difference between a conversation where the flow of information is one way (monologue) and a conversation where the flow is both ways (dialogue). The constant invocation of dialogue as the mantra that will save nanotechnology from the disastrous fate of biotechnology and the ignoring of the double-sidedness of the concept of dialogue is perhaps the most striking similarity between the two technologies. As discussed by Nielsen *et al.*, the idea of public participation in forming the policies for emerging technologies such as biotechnology and nanotechnology is very widely accepted today. But whereas the general idea of public participation and the concept of dialogue are almost universal within the Western world (if one can actually say that), the contents of these notions are so far from being self-evident that they lean upon the obscure (Nielsen *et al.*, 2004).

The content of the idea of public participation can very generally speaking be said to be decided by the reasons for supporting the idea in the first place. There seems to be a continuum from, on one hand, those that support public participation because it is a way of ensuring the public some sort of democratic or semi-democratic influence on the way that new technologies are supported through research-funding and in the application phase to, on the other hand, those who see public participation as a way of merely legitimizing the technologies in the eyes of the public. In the first case the goal is to live up to some democratic ideals of some sort without influencing the result of the participation. In the other case the whole point is to get the technologies accepted.

In the real world the motivation for seeking public participation is seldom a clear-cut case, placing the motivation somewhere between the two extremes in the continuum. But as a rule of thumb I guess it would be fair to say that the closer one is to be motivated by democratic ideals rather than being motivated by interest in having the technologies accepted, the more content of a meaningful kind can be placed in the concept of dialogue.

As mentioned above, this has most clearly been seen in the notion of the knowledge deficit model. Although this has been refuted both with regards to biotechnology and (although with less empirical material) nanotechnology (The Royal Society, 2004; The Danish Board on Technology, 2004; Cobb and Macoubrie, 2004) there is still a belief that the concept of dialogue can be transformed into a monological information stream that will result in wider acceptance of a given technology.

At the same time another extreme form of dialogue seems to be forming; a dialogue that is as monological as the first, but that has the public rather than the scientists doing all the talking. It consists of different stakeholders (especially policymakers/industry) believing that the way to have a dialogue is to ask the other person what he or she wants, and then give it to him or her. So when facing technological developments where it is not foreseeable how the public will react, the way to use dialogue is to ask the public (typically through quantitative polling) what kind of development they want and then try to bring that development about.

And what, you may ask, is the problem with that? I have just lamented that missing dialogue in the first extreme because it concentrates purely on the acceptance aspect. Why not welcome this attentive listening to the opinion of the public? The reason is that it turns dialogue into marketing research. The preferred level of the public's knowledge on the subject can always be discussed. Should they just be polled about their top-of-the-head opinion? Or should they be offered some kind of chance to actually deliberate the questions in hand. But in the end it will still just be marketing research, done more or less eloquently and in a sophisticated way. One thing is for certain—as long as the answers that one obtains from such research are not thematized, analyzed, and evaluated in some kind of critical discussion, but are just used as weathercocks—ways to figure what to do and especially what not to do—the word dialogue remains just as empty as in the first extreme.

Dialogue is a complicated concept with a long and entangled history within the disciplines of philosophy and ethics. Without wanting to end up in a habermasian discourse—ethics where the description of the ideal dialogue somehow ends up answering the ethical questions that the dialogue originally was a method of agreeing and disagreeing about—I believe it to be very important that the concept of dialogue that lies behind different ways of envisioning public participation in relation to technologies are made very clear in every discussion of "how to do it" in this area. As always, attempts to answer the "How" question always bring the "Why" question into question!

Basically, dialogue is a way of ethically balancing two very important considerations: respect for the other person, whether it is one or "the public," and taking responsibility for one's own views of the world and attempting to do what one sincerely believes to be in the best interest of the other (whoever that may be). A classical example of this conflict can be found in the relationship between the physician and the patient. The ethical duty for the physician is to balance between taking all responsibility from the patient (paternalism as tyranny) and just leaving the patient on his or her own in trying to decide upon different methods of treatment in the mistaken belief that the task of a physician is just to provide neutral information, respecting the patient's right to self-determinacy at all costs (autonomy as denial of responsibility).

The concept of dialogue implies that there are two or more different opinions about something and that the people holding these opinions are willing to discuss them, holding a small window open in the back of their minds to the possibility that they may be wrong. A dialogue in which it is decided from the outset that only one part of the dialogue (and that is usually the other part) could end up changing their minds is no dialogue, but could better be described as a caricature of energetic and zealous religious proselytizing.

I readily admit that these remarks give no concrete suggestions to how the concept of dialogue should be operationalized, if the aim is to have a "true" dialogue. I will just point out that realism in policymaking is a very good thing. There is no need to fly off the cliff when we all know the political reality, but when engaging in philosophy and ethics it might just be necessary to take the jump and begin by leaving political considerations aside and visualizing the ultimate goals that humans could choose to seek. Patterson (1998, p. 148) expressed such an ideal:

> Dialogue does not mean simply an exchange of views between two parties, eyeball to eyeball in the conference chamber. In its fullest sense it is what the early Romantics called "sym-philosophy," a term I should like loosely to translate as "lovingly seeking wisdom together."

The conclusion to these remarks is thus rather simple. Before entering into dialogue and figuring out ways to engage the public in such an endeavor it is (from an ethical point of view) paramount that we (and who this "we" is, is indeed a huge question in itself) ask the "Why" question. Our answers to that will to a large degree answer the "How" question too. And we should perhaps begin to do this. The possibility that hype-concepts such as "dialogue" and "public participation" could end up only bringing a slight yawn as reaction to an invitation becomes more realistic the more they are seen as just that: empty concepts used as rhetorical devices by politicians and scientists whom "the public" has perhaps never trusted as little as they do these days.

Thus the third teaching to be taken home from the biotechnology experience is that dialogues are complicated, time-consuming endeavors with unpredictable outcomes. And that the reduction of dialogue into monological information is not only philosophically unsound and wrong from both an ethical and a strategic point of view, but also a certain way of killing what public enthusiasm might be left out there.

Conclusions

"What can nanotechnology learn from biotechnology?" I hope to have shown you here that the answer to this question depends on why it is asked and who is asking it. I have sought to highlight three teachings that I think are important from the biotechnology experience that we might as well use in the discussions about the ethical and societal concerns about nanotechnology. These teachings concern the importance of knowing what is discussed, whether disagreements are about factual stuff or values and whether the concept of dialogue is used correctly when all it points to is an "information" campaign.

Are these the most important lessons to be had? Again, it all depends on the perspective from which you see the emerging debate, and especially the outcome you hope for. In this essay I have tried to see things from the perspective of society. Not society understood as an economic entity or an entity in need of technological development, but society understood as a gathering of individuals who have to figure out together how to proceed into the future. A future where it is most likely that nanotechnology will play a growing role. It is therefore crucial that we figure out ways in which to discuss what kind of future we would like to live in and how to bring it about. That is, as I see it, the only ethical way to create socially robust methods of integrating nanotechnology into our societies.

Endnotes

1. It is an interesting phenomenon how the case of GM crops in Europe is more than frequently referred to among scientists from both natural science and the humanities (including ethics!) as a disaster, catastrophe, problem, or just plainly something to be avoided. Seen from the perspective of the citizens (and people endorsing the free market) it could just as well be seen as a huge success where an unwanted technology failed to gain a foothold in the market, thus proving that the idea of demand and supply actually works.
2. Some of Løgstrup's writings have been translated into English (see Løgstrup, 1995, 1997).
3. I will get back to this common misunderstanding in the next section.
4. Cf. Deficit model as used by Macnaghten and Priest in Chapters 6 and 11.
5. The policy layer at the second level is perhaps not so much a layer in itself as it is the place where we discuss both specific applications and visions of nanotechnological utopias and dystopias.
6. My agreement with Bainbridge about the two distinct layers does not extend to his understanding of who is to be found within the two layers. As nanotechnology continues to be the most overhyped technology of all times (Berube, 2005), there can be no doubt that scientists, and especially research-funding agencies, are engaged in science-fiction telling to attract funding and justify it to the public. The idea that scientists are somehow not involved in the hype of nanotechnology overlooks the very simple fact that they are usually competing about the grants and the public's attention. But as Berube

also points out, the most hysterical examples of overselling can be found in governmental reports that seem to suppose that if the words "revolutionize," "economic growth," and "societal benefits" are mentioned often enough, these things will occur. See Berube (2005, pp. 155–184) for a series of examples. My two personal favorites continue to be: *Nanotechnology: Shaping the World Atom by Atom* by the US National Science and Technology Council from 1999 and Roco and Bainbridge (eds) *Societal Implications of Nanoscience and Nanotechnology* from 2001.

References

Bainbridge, W. S. (2002). Public Attitudes towards Biotechnology. *J Nanoparticle Res* **4**, 561–570.

Berube, D. M. (2005). *Nano-Hype: The Truth Behind the Nanotechnology Buzz*. Prometheus Books.

Cobb, M. D. and Macoubrie, J. (2004). Public perceptions about nano-technology. *J Nanoparticle Res* **6**, 395–405.

Cooper, D. E. (1998). Intervention, humility and animal integrity. In: Holland, A. and Johnson, A., ed. *Animal Biotechnology and Ethics*. Chapman & Hall, pp. 145–155.

Danish Board on Technology (2004). *Public Views on Nanotechnology*. Copenhagen.

Dawkins, R. (1998). *Unweaving the Rainbow*. Penguin Press.

Grinbaum, A. and Dupuy, J. P. (2004). Living with uncertainty. *Techné: Res Philos Technol* **8**, 4–25.

Jensen, O. (2001). At hente rummet ind igen. Teologiske betragtninger over vort naturforhold. In: Madsen, L. D. and Gjerris, M., eds. *Naturens sande betydhning—om natursyn, etik og teologi*. Multivers, pp. 78–105.

Kunic, J. C. (2003). *The Naked Clone: How Cloning Bans Threaten Our Personal Rights*. Praeger Publishers.

Lassen, J., Gjerris, M., and Sandøe, P. (2005). After Dolly—ethical limits to the use of biotechnology on farm animals. *Theriogenology* **65**, 992–1004.

Løgstrup, K. E. (1995). *Metaphysics*, Vols I–II. Marquette University.

Løgstrup, K. E. (1997). *The Ethical Demand*. University of Notre Dame Press.

National Science and Technology Council (1999). *Nanotechnology: Shaping the World Atom by Atom.* National Science and Technology Council.

Nielsen A. P., Lassen, J., and Sandøe, P. (2004). Involving the public: participatory methods and democratic ideals. *Global Bioethics* **17**, 191–201.

Patterson, G. (1998). *The End of Theology: And the Task of Thinking about God.* SCM Press Ltd.

Peters, T. (1996). *Playing God: Genetic Determinism and Human Freedom.* Routledge.

Reiss, M. J. and Straughan, R. (2000). *Improving Nature? The Science and Ethics of Genetic Engineering.* Cambridge University Press.

Roco, M. C. and Bainbridge, W. S. (eds) (2001). *Societal Implications of Nanoscience and Nanotechnology.* Kluwer Academic.

Royal Society (2004). *Nanotechnology: Views of the Public. Quantitative and Qualitative Research Carried out as Part of the Nanotechnology Study.* The Royal Society.

Royal Society and Royal Academy of Engineering (2004). *Nanoscience and Nanotechnologies. Opportunities and Uncertainties.* The Royal Society.

Sandler, R. and Kay, W. D. (2006). The GMO-nanotech (dis)analogy? *Bull Sci Technol Soc* **26**, 57–62.

Internet references

Buerger, M. E. (2006). *From the Enlightenment to N-lightenment.* http://wise-nano.org/w/Buerger_CTF_Essay

European Commission (2005). *Nanosciences and Nanotechnologies: An Action Plan for Europe 2005–2009.* Communication from the Commission to the Council, the European Parliament and the Economic and Social Committee. European Commission. http://europa.eu.int/comm/research/industrial_technologies/pdf/nano_action_plan_en.pdf

Gjerris, M. (2006). Ethics and farm animal cloning: risk, values and conflicts. The Danish Centre for Bioethics and Risk Assessment. http://www.sl.kvl.dk/cloninginpublic

Lassen, J. (2005). Public perceptions of farm animal cloning in Europe. The Danish Centre for Bioethics and Risk Assessment. http://www.sl.kvl.dk/cloninginpublic

Meyer, G. (2005). Why clone farm animals? Goals, motives, assumptions, values and concerns among European scientists working with cloning of farm animals. Project report 8, Danish Centre for Bioethics and Risk Assessment. http://www.sl.kvl.dk/cloningin public

Nordmann, A. (2004). Converging technologies: shaping the future of European societies. European Communities. http://europa.eu.int/comm/research/conferences/2004/ntw/pdf/final_report_en.pdf

Roco, M. C. and Bainbridge, W. S. (eds) (2002). *Converging Technologies for Improving Human Performance. Nanotechnology, Biotechnology, Information Technology and Cognitive Science.* Report sponsored by the National Science Foundation, USA. http://www.wtec.org/ConvergingTechnologies/Report/NBIC_report.pdf

From Bio to Nano: Learning the Lessons, Interrogating the Comparisons[1]

6

Philip Macnaghten

Introduction

Given the starkness of the "GM controversy," particularly as it unfolded in Europe, it is not surprising there has been speculation as to whether nanotechnologies might experience a similarly rough passage. Here is another potentially transformative technology, subject to similar levels of utopian promise, expectation, and dystopian fear (Nordmann, 2004). Crudely put, the GM experience represents a warning, a cautionary tale of how not to allay public concern. Avoiding nanotechnology becoming "the next GM?" is

What Can Nanotechnology Learn from Biotechnology?
ISBN: 978-012-373990-2

seen as critical to the public acceptability of applications in the field (see, for example, Mayer, 2002; Wolfson, 2003; Einsiedel and Goldenberg, 2004; Mehta, 2004).

Under scrutiny, however, the GM–nano analogy quickly breaks down. These are very different technical endeavors, emanating from different disciplines. One is a particular type of application, the other a catch-all for a multitude of products and processes. So a direct comparison between them may be of limited value. We agree with the authors of one recent paper that the analogy "is not as strong or as helpful as its ubiquity would suggest … [and] therefore needs to be employed advisedly" (Sandler and Kay, 2006).

But as we have suggested elsewhere, the GM case can still be useful to illustrate how policymakers struggle to handle emerging technologies in the early stage of their development (Wilsdon and Willis, 2004; Macnaghten *et al.*, 2005). There are also various ways in which the GM experience has shaped, and will continue to shape, political, regulatory, and public debates around nanotechnologies.

This chapter offers some critical reflection on two recent pieces of empirical material: on a series of interviews with key individuals active in the pre-1999 debates over GM plants and crops in Europe;[2] and on a set of public focus group discussions on emergent public attitudes towards nanotechnologies.[3] We begin with the interview data.

Learning from the past

Competing understandings of "the science"

In the 1970s many leading genetic scientists expressed effusive visions of the transformative societal futures that would result from advances in genetics and biology. One such figure, C. H. Waddington, described the arrival of genetics as presaging a "second industrial revolution," which would overturn the destructive effects of the first revolution, which was based (in his view) on physics and chemistry (Waddington, 1978). Visions such as Waddington's were not simply scientific imaginaries. They were social too.

One of our interviewees, Professor Nigel Poole, articulated such an imaginary when he spoke with passion about the potential for genetic and plant science to transform the economy:

> I remember so clearly getting a very passionate talk, a lecture, evangelical almost about the future of biotech. This must have been in the

very early 1970s. And I was totally convinced—that in biotech we would start to see the end of the chemical industry or massive change in the chemical industry. And I think they even said that by the turn of the millennium the chemical industry would have been gone. ... I don't really think then we were thinking about DNA, you know gene therapy and that stuff—that was a bit too early. But those were the dreams and that's still my belief. It's a belief that goes right back to 1972. (N. Poole, March 16, 2004, personal communication)

In the commercial sphere, Monsanto's initial R&D commitment to GM crops was justified in terms of equally positive visions for the future of global agriculture, beyond more technical visions of "terminator technology" or proprietary brand herbicide-resistance (Doubleday, 2004). Although now often disparaged as having been focused exclusively on corporate profit and control, Monsanto's imaginaries in the 1980s and 1990s reflected a vision of a more environmentally benign system of food production. Equally striking, however, was the degree of naïveté within this vision about other actors' responses and expectations.

Societal and scientific imaginaries of this kind—projections of future imagined worlds—frequently inform and shape new scientific fields. The GM experience points to the fact that, despite their scientific significance and persuasive power for governments and investors, such imaginaries tend to be insulated from wider recognition, accountability, and negotiation. They are shielded by myths about the purity of science and assumptions of a linear relationship between scientific research and the public domain. (Wynne, 1995) According to this model, it is only when scientific knowledge is thought to have potential "applications" that social and ethical dimensions enter in. This means that social issues are acknowledged to arise only in connection with possible impacts, not with the aims and purposes underlying the production of scientific knowledge.

In the last decade or more, however, this model has come under increasingly intense pressure, partly due to the changing political economy of research where commercial exploitation and property rights have become central, and partly due to the emerging policy significance of "public engagement" in the UK and EU. Under these conditions, the need for even "basic" scientists to project images of how their research might benefit society in the future, has intensified. As basic research comes to be called "pre-market" research, an unavoidable implication is that "basic" research practices are imagining possible market outcomes, in ways that may subtly but significantly shape those research agendas and cultures themselves.

The limits of risk assessment

The regulatory context for GM crops was framed by a particular conception of risk assessment; one which was methodologically quantitative and almost exclusively concerned with the "direct" effects of individual crops. Wider questions arising from the overall social, ecological, medical, and political implications of GM technology were marginal to official considerations. This limited framework of risk assessment, coupled with official assurances of safety, had the effect of making the official mechanism for risk assessment a de facto locus for the political contestation of GM releases (Ravetz, 2005). It also played a role in the formation of public controversy in the late 1990s.

In the UK, the Environmental Protection Act 1990 established the Advisory Committee for Releases to the Environment (ACRE) as the formal body responsible for assessing the risks to human health and the environment from the release of GM organisms. ACRE's position was awkward from the outset. As the only established mechanism for the regulatory assessment of GM releases this advisory body became the de facto political authority on GM releases, backed by the UK Government's commitment to "sound science" (Mayer and Stirling, 2002). However, ACRE was concerned solely with the risks of individual GM crops. In seeking to address specific risks on a case-by-case basis, this risk assessment template came to be structurally built on past knowledge, rather than taking account of the potential for new types of hazards that might arise in unknown forms (Tait and Levidow, 1992).

The ex-Chair of ACRE, John Berringer, confirmed in his interview with us the difficulties that this methodology created in relation to the wider cumulative implications of GM crops:

> We recognized quite quickly in ACRE that it was really very easy to give approval, say, for GM maize as is being done at the moment. You could not see any human risks, you couldn't really see any serious environmental ones, and as was proven in the farm trials, it's actually slightly better than traditional herbicide treatment in terms of wildlife. But we asked the question, sure, we can do this for one crop, one manipulation. But when all crops are being manipulated, every affect becomes additive. So if you approve an insect-resistant oilseed rape, you can do an analysis and say well, that particular variety is only likely to occupy such a percentage of the area of the UK. The impact on insect production is small, the impact on birds is therefore likely to be small, probably quite acceptable. … However, if every farmer grew those crops at every farm, suddenly the impact is enormous. Where is the mechanism to put it all together? (J. Berringer, March 23, 2004, personal communication)

He expanded on this concern later in the interview:

> The big issue in terms of commercialising is what happens if you then approve another variety with another gene and then another variety with another gene. You'd need to know something about the inter-relationship of those genes if they come together. And I finished chairing the committee before it was properly decided. ... First person's dead easy, second person has to take into consideration the first gene, the third has to take into consideration the first two, the fourth has then got three prior genes plus their own. So there were lots of arguments. I think it's still not remotely solved as to what happens when you've got lots of different genes out there. (J. Berringer, March 23, 2004, personal communication)

Though initially imagined in precautionary terms, ACRE's reductionist framing stunted the extent to which real-world contingencies could be thoroughly considered. This led to mounting problems for the authorities responsible for the regulation of biotechnologies. Importantly, the limited framework of risk assessment was also intimately linked to the marginalization of wider social and ethical concerns about GM food (Jasanoff, 2005). Such concerns—including the perception that government decisions had already been taken, that GM foods would lead to an inevitable diminution in consumer choice, of GM as unnatural, and concerns about corporate control of food systems—were simply not captured by the language of risk and safety (Grove-White *et al.*, 1997).

The effect of this deletion was to make debates about the risk and safety of GM crops stand-in for a host of other unacknowledged concerns (Gaskell, 2004). Yet the poignancy of these wider social concerns was redoubled by the lack of any official recognition and official assurance of the adequacy of assessment mechanisms. And for these precise reasons, ACRE became the de facto locus for the political contestation of GM releases.

Other European governments, including Denmark, the Netherlands, Germany, and Norway, responded to the concerns raised about GM with more innovative forms of social debate and dialogue. Building on these, two such initiatives were undertaken in the UK—a consensus conference, organized in 1995 by the Science Museum and AFRC, and a government-organized "National Biotechnology Conference" held in early 1997 (Joss and Durant, 1995; Macrory, 1997). Unfortunately, both of these initiatives were limited in their scope, public visibility, and ability to shape the trajectory of GM regulation and development. Similarly neither was framed to enable detailed examination of wider societal and ethical concerns.

Competing understandings of "the public"

During the 1970s and 1980s, public attitudes to nuclear power were systematically characterized as subjective, emotional, and false risk perceptions (Wynne, 1992). In the early 1990s, an equivalent dynamic emerged in the biotechnology field. With a few exceptions, it was assumed that public concerns about GM crops could only be founded on an incorrect understanding of the technology or a complete lack of knowledge altogether.

As the 1990s advanced, social science researchers became increasingly active observers of the state of public opinion in relation to GM plants and foods (INRA, 1993, 2000; Durant *et al.*, 1998; MORI, 1999). Much of this work focused on public attitudes rather than underlying sources of social tension, and how these reflected limitations in the risk-regulatory framework itself. Indeed, most built on the assumption that the discourse of atomized science-defined "risks" offered an analytically sound basis for commentary on the state of public opinion. As such, even though survey data began to point to a steady decline of public confidence towards biotechnology throughout the 1990s, this provided little explanation or warning for why GM would become the focus of such controversy. The assumption was that the key issues of public concern were the risks as defined by risk assessment, and that any disinclination by the public to accept such risks was based on a (false) belief that the risks were too high.

Even following the official discrediting of this "deficit model" (symbolically put to bed in the House of Lords Science and Society report in 2000), this misconception continues to be resurrected, albeit in a succession of new versions. Such persistence reflects an institutional science and policy culture which continues to project problems of public conflict, mistrust, and skepticism about prevailing science onto other supposedly blameworthy agents—often a sensationalist media or mischievous NGOs. Responsibility for such problems is continually externalized away from official institutions, such that government and scientists' own roles are rarely questioned.

Some of our interviewees reflected this view:

> There was a clear view that there was an anti-science agenda that was coming through. … The biggest frustration was the dishonesty and the distortion [on the part of NGOs and the media] which it's very difficult to handle. It's extraordinarily difficult to handle. (R. Baker, February 24, 2004, personal communication)

Fear of the unknown … it's like MMR in many ways. You know, no real benefit—and fear of the consequences—and a confusion because they were being fed downright lies by people. There is no way of actually correcting the [NGO] lies. (N. Poole, March 16, 2004, personal communication)

The implication is that NGOs purposefully acted to manipulate policy and create controversy. Yet interviews with NGO actors involved with GM campaigns throughout the 1980s and 1990s suggest that such charges misrepresent the capacity of NGOs. National bodies like Greenpeace, Friends of the Earth, the Royal Society for the Protection of Birds (RSPB), and the Soil Association, each of which made a distinct contribution to the more visible stages of the GM controversy, tend to be preoccupied with multiple issues. In the UK, these NGOs were relatively slow and uneven in developing coherent campaigns on GM crops. Indeed the overall response by these groups to GM lacked clarity and unanimity. Greenpeace for example, following its initial direct action drawing attention to Monsanto's first ship import of GM soya in mid-1996, was uncertain what to do next. There was protracted internal discussion within the UK office about whether there was any appropriate basis for further initiatives. Friends of the Earth took the issue up only in 1998, in parallel with the RSPB's shared concern over the specific issue of potential biodiversity impacts from commercial growing of GM crops. This led to the setting-up of the Government's farm-scale trials at the end of that year.

So far from leading the mounting controversies about GM commercialization up to this point, the NGOs found themselves in the position of responding to the intensity of wider public unease being expressed through the spontaneous emergence of new networks and initiatives.[4] Whatever the beliefs or inclinations of individual supporters or staff members, NGOs face constraints in their ability to influence or transmit the full range of concerns of the wider population in relation to new technologies. Much of the difficulty for Greenpeace, Friends of the Earth, and others in campaigning coherently on GM-related issues arose from the fact that the dominant "risk" discourse offered them minimal scope for interventions. For example Greenpeace's stated approach to GM issues was articulated in the idioms of science alone:

> The difficulty Greenpeace has, is that we are a global organization and, if one is to take value-based stances on what is and is not natural and the value judgments and the sort of loadings that that comes with, how relevant is it to talk about it in those terms and try and explain one's concern in those terms in China, where the term for nature doesn't actually exist or certainly doesn't exist in any meaningful form that we would recognize in the West? ... That is not our position.

Our position is about scientific risks. Our kind of globally applicable standard is the science of environmental risk. You can say that's the basis of our campaign policy and that's where we're coming from. (D. Parr, March 4, 2004, personal communication)

Condensation points

By the end of the 1990s, GM crops had become something of an iconic environmental and social issue in many countries. At the immediate level, concern crystallized around the potential for unforeseen ecological consequences and the implications of GM for agriculture and food production. But discussion of the technology also reflected a broader set of tensions: global drives towards new forms of proprietary knowledge; shifting patterns of ownership and control in the food chain; issues of corporate responsibility and corporate proximity to governments; intensifying relationships of science to the worlds of power and commerce; unease about hubristic approaches to limits in human understanding; conflicting interpretations of what might be meant by sustainable development. These and numerous other "non-scientific" issues condensed onto GM crops because of a particular range of institutional and cultural contingencies shaping the technology and its development (Grove-White *et al.*, 2000; Wynne, 2001).

This was hardly without precedent. In the very different circumstances of the 1970s, disputes about civil nuclear power had played something of an analogous role. Here too was an apparently unstoppable technology that became a vector for both issue-specific concerns and more general social and political anxieties. Beyond detailed challenges about nuclear safety and open-ended problems of nuclear wastes, wider issues presented themselves in intense forms. For both GM and nuclear power, these arguments reflected not simply "technical" issues held to be legitimate by governments and scientists, but also wider social relations in which the respective technologies were embedded. In the absence of other meaningful spaces in which such debates could take place, GM became the occasion and the opportunity.

Learning from the present

We turn now to the present and specifically to the learning to be drawn from the ways in which lay people's experience of GM, and of its handling by key institutions, appeared to be shaping emergent public attitudes of nanotechnologies. We report on the ways that GM acted as a heuristic in the group discussions as providing grounds to suspect that nanotechnologies might have future disruptive effects and pose considerable problems for governance.

I don't feel I have any control

At the beginning of the group discussions people were asked to discuss their perceptions of technology and to imagine their likely social impacts in the future. Not unsurprisingly people voiced both enthusiasm and disquiet. The anticipated pace, scope, and intensity of technological change were the source of considerable concern. More surprising was the iconic role attached to the new genetics, and particularly the case of GM foods. People's experience of GM was of a technology imposed upon them and with associated and potentially disruptive social impacts. Across the discussions people discussed GM as paradigmatic of the likely social impacts of future technological change: technologies which people would have little power to affect, where ownership and control would be further consolidated into large and unaccountable actors, outside the reach of citizens and national governments, and where the impacts would be unpredictable, disruptive, and potentially uncontrollable. Below is one such extract from a group of professional women discussing the undemocratic dimensions of genetic technologies:

> F: Well, things like genetics … It would be interesting to see how it plays out. But I don't feel that I have control over [or] any input in to how that happens, you know like cloning or genetic modification …, it's rushing very quickly ahead. I don't ever feel like that's been an election issue or been in someone's manifesto. These sorts of things I think are going to be really big questions for humanity and I think that they're not really on [anyone's] agenda; but I don't feel that there that there is any way that I can express my opinion. (Professional women discussing nanotechnology, 2005)

We have not had a say again

A further dimension of people's discomfort with GM was the sensed inadequacies of the political system to address the ethical, social and health implications of the technology in advance of its application. Again people's experience of GM was seen as providing grounds for the need to adopt a more cautious approach to future technological innovation in the round:

> **F1:** What you're saying is we haven't had a say again. In that these things are just coming through and ...
> **F2:** They don't feel the need, no
> **F1:** But also the speed with which things are going forward as well, like I was trying to say before, I don't know there are a lot of well-publicized questions around genetic modification which ... I don't feel have been addressed, ethical question haven't been addressed really or publicly ... I'm a little bit wary about jumping into, rushing forward with another new technology where I feel that the old questions haven't even been addressed. (Professional women discussing nanotechnology, 2005)

We'll be the guinea pigs

When people began to discuss the issues posed by nanotechnologies we were struck by striking parallels between emergent attitudes and public attitude research on biotechnologies in food conducted in 1997 at a time at which attitudes were similarly latent (R. Baker, February 24, 2004, personal communication). Across both cases our public participants expressed ambivalence as to whether they would have much of a say in the direction and pace at which the technology developed; whether control and ownership of the technology would be further consolidated in the hands of the few; whether governments would be able to address the ethical, social and health implications of the technology in advance of their application; and more broadly about the apparent hubris of the visions and imaginaries of the technology promoted by proponents. Parallel textures of concern are evidenced below:

> **F1:** It's amazing.
> **F2:** I find it quite daunting actually, I find it a bit scary.
> **F1:** This is the vision of the robotic environment with everything controlled for you and everything 100% perfect and plastic.
> **F2:** It's like even the food ... You buy a piece of fruit, it's healthy, after a piece of time it wrinkles, you throw it away or whatever and that is a natural process and I think in some ways it's kind of fiddling with that natural process.
> **F3:** It's like trying to make a perfect race again.
> **F4:** We just don't know the long-term effects do we? That's the problem.
> **F2:** So basically our generation's going to be the ones that they test this all out on. If it all goes horribly wrong, we'll be the guinea pigs. (Mothers discussing nanotechnology, 2005)

F1: I started out not too bad when I had the discussion. I thought I'd have an open mind about it, but I've changed my mind. As soon as I saw that about the human gene, suddenly the enormity of it made me feel really awful. I got an awful feeling about it, because I thought it was something that ... I think we're touching things that we don't realize and I think that we're taking things out of the earth, and we're now trying to correct it by using things like genetic engineering, because mistakes were made. And I feel that time's just ticking by and we don't realize what's going to happen in the future. I think something terrible could happen. It's given me a bad feeling really.

F2: It's a frightening thought to think that time's ticking away though.

F1: Yes. It's something that I'd like to put at the back of my mind now. I wouldn't like to think about it again. I probably wouldn't—but when we talk about it, it does bring it to your mind. But then I'll probably put it to the back of my mind. (Working women discussing biotechnology, 1997)

Dark scary futures

However, there were also notable differences. Whereas biotechnology appeared a relatively tangible and graspable concept, nanotechnology appeared more opaque to our participants. Defined simply by scale people struggled to develop a collective imagination of what nanotechnology was and of the ways in which it would impinge on everyday life. Some application domains were seen as exciting, necessary, and beneficial, especially in the medical domain. Nanotechnology clearly had a "wow" factor that had not been present in analogous discussions of biotechnologies.

A further difference was the sheer density of issues attached to nanotechnology. In addition to those set out above, nanotechnologies were seen to have the potential to transgress boundaries between human and machine, to facilitate new forms of control and surveillance, and to intervene at an even more fundamental level on natural processes with unknown consequences. Our participants worked hard to develop an understanding of how such processes might be adopted by powerful actors, including governments, corporations and the military, and began to entertain "some dark scary futures:"

M1: I think the worrying thing for me ... is that it's almost as though we lose control of what's going on because the technology itself is capable of replicating and you know pretty much making its own decisions.

M2: I think that is a big problem. It's like the thing you were saying with creativity as well. If the human controls the technology that's fine, as soon as it becomes the technology making all the decisions then that's when you have a problem because humans are completely different from a computer.

M1: There are some scary dark futures where you have strains of children who are and are not enhanced in some way, and that's a really dodgy thing.

M3: Do you have your kids injected at birth to enhance the way their muscles grow and things. (Technology group discussing nanotechnology, 2005)

It must be an absolute godsend to the terrorists

An equally potent and apparently novel dynamic was the capability for nanotechnologies to be used for purposes not imagined by their original developers, especially in the new security environment. Several participants saw nanotechnology potentially being used by terrorists. This dynamic clearly added to the controversy potential of the technology:

M1: The more I think of the dangers, the more evil applications I can think of using nanotechnology.

M2: Well I just find it quite frightening really. I think it's quite disturbing. The potential to harm seems to me to be greater than the potential for good if it gets into the wrong hands. (Technology group discussing nanotechnology, 2005)

Int: How controversial do you think it's going to be?

M1: Far more than genetic modification.

M2: It's going to be more. And what are the fault-lines through which it's going to become politically controversial?

M1: The medical, the human biological angles as well as the food chain.

M3: I would have thought in the present climate particularly terrorism. It must be an absolute godsend to the terrorists this sort of technology. (Professional men discussing nanotechnology, 2005)

Why can't we slow it down?

All of these factors contributed to the perceived difficulty of imagining robust and effective systems of governance and regulation. It was generally seen as unrealistic to advocate a slower, more cautious approach to nanotechnologies. Some suggested that an overly precautionary approach could harm the UK's economy and lead to outward investment. Others observed that much innovation is transnational and increasingly beyond the control of individual governments. It was widely felt that the pressure for commercial return would lead to corners being cut.

F1: The whole thing we've been talking about is that these things happen so quickly, why can't we slow it down? Is it going to matter that much if it is slowed down?

F2: But say this country does that and slows it down then you're gonna go abroad ... yeah, and it's gonna come back into this country anyway. (Professional women discussing nanotechnology, 2005)

Lessons for nanotechnologies

So what implications can we draw from this account for future approaches to nanotechnologies? First, when faced with new situations and technologies, regulators will usually turn to assessment frameworks developed for previous technologies and tied into existing debates. Given this tendency to "fight the last war," there is a need for more textured, socially realistic analysis of the distinctive character of particular technologies, and greater recognition of the limitations of conventional models of risk assessment.

Second, it is important to be more realistic about the diverse roles of NGOs. The breadth and unfamiliarity of issues now being thrown up by new technologies mean that NGO responses are in continuing flux, and a richer account of the ways in which NGOs "represent" opinion in wider society is needed.

Third, the GM case suggests that the deficit model of public skepticism or mistrust of science and technology is a fundamental obstacle for institutions charged with the regulation and assessment of new technologies. For nanotechnologies, there is a need to build in more complex and mature models of publics into "upstream" policies and practices.

Fourth, GM demonstrates the ways in which new technologies often operate as nodal points around which wider public concerns condense. Such processes of condensation are inherently unpredictable. However, a richer understanding of the underlying dynamics of such processes—informed by recent thinking in the social sciences—could begin to provide some clues. In considering approaches to the social handling of nanotechnology and its potential manifestations in applied forms, care will need to be taken to "design in" greater social resilience.

Finally, our research on emerging public attitudes suggests clear parallels between nano and bio. In both cases people expressed ambivalence, fatalism, anxiety over the directions in which the technology was moving, as well as skepticism in the ability of governments and regulators to exercise adequate oversight. In both cases this ambivalence did not diminish through greater knowledge and awareness. Instead, through exposure to the multiple ways in which the debate was being characterized, and through debate and deliberation, our participants moved towards a more skeptical view as to the ability of government and industry to represent the public interest.

However, while harboring unease, participants also saw the considerable promise for nanotechnologies to contribute to the social good. While the social visions tacit in GM were never openly acknowledged or subjected to public discussion, this remains still an opportunity for nanotechnology. In essence, a more open model of innovation is required, in which imaginaries are opened up to greater scrutiny and debate. We will need to open up the "black boxes" of science and innovation, to induce greater reflexive awareness amongst scientists, policymakers, corporations, and others. In this way, innovation processes may indirectly gain added sensitivity to diverse human needs and aspirations, and so achieve greater resilience and sustainability.

Endnotes

1. This paper emerges from a genuinely collaborative research effort involving researchers from Lancaster University and Demos. The author would like to thank colleagues Matthew Kearnes, Robin Grove-White, James Wilsdon and Brian Wynne for their contribution to the research and in particular to an earlier version of this paper. Responsibility for this paper remains with the author. The research project, titled 'Nanotechnology, Risk and Sustainability: Moving Public Engagement Upstream' was funded by the Economic and Social Research Council (RES-338-25-0006).
2. The principal focus of our analysis is on the 1980s and 1990s, up to the moment when the controversies over the first period of GM development reached their peak in the UK, in February 1999. Clearly, since that time, there have been a number of further developments, including the creation of the Agriculture and Environment Biotechnology Commission (AEBC), the UK Government's GM Dialogue, completion of the farm-scale trials, and, not least, the hearings at the World Trade Organization (WTO) into the formal US

complaint on "Biotech Products" against the European Union. But we have drawn the line at February 1999, in order to reflect on the underlying processes which shaped the controversies, rather than the unfolding of the post-1998 events themselves.

3. The purpose of the focus groups was to encourage discussion of potential issues arising for nanotechnology, within a framework set by participants rather than imposed by official regulatory and risk-assessment vocabularies. The sample consisted of five groups, each of which met twice, with a gap of one week between the sessions. Participants were recruited on the basis of their existing participation in local community or political issues, but with no prior involvement or exposure to nanotechnology. They included a group of professional men (doctors, architects, civil servants, etc.); a group of professional women (mostly employed as middle managers in business); a mixed group with demonstrable political interests; a group of mothers with children of school age; and a mixed group with an interest in technology. The groups were conducted in Manchester and London.

4. Indeed, as a response to the perception that such groups were not campaigning actively on GMOs from the mid-1990s, wider bodies of opinion, independent of such organizations, crystallized in a host of more ad hoc and GM-specific networks—including Genetix Snowball, the Genetics Network, the Genetics Alliance, Corporate Watch, Genewatch, and many others. This further range of frequently Internet-focused associations embraced wide and diverse constituencies of concern, and can be read as "organizational" crystallizations of the pervasive, but previously latent, public unease about GM-related issues noted in UK social research as early as 1996–1997 (Grove-White *et al.*, 1997).

References

Doubleday, R. (2004). Political innovation: corporate engagements in controversy over genetically modified foods. Unpublished PhD thesis, University College London.

Durant, J., Bauer, M., and Gaskell, G. (1998). *Biotechnology in the Public Sphere: A European Source Book*. Science Museum.

Einsiedel, E. F. and Goldenberg, L. (2004). Dwarfing the social? nanotechnology lessons from the biotechnology front. *Bull Sci Technol Soc* **24**, 28–33.

Gaskell, G. (2004). GM foods and the misperception of risk perception. *Risk Anal* **24**, 185–194.

Grove-White, R., Macnaghten, P., Mayer, S., and Wynne, B. (1997). *Uncertain World: Genetically Modified Organisms, Food and Public Attitudes in Britain*. IEPPP, Lancaster University.

Grove-White, R., Macnaghten, P., and Wynne, B. (2000). *Wising Up: The Public and New Technologies*. IEPPP, Lancaster University.

INRA, Institut National de la Recherche Agronomique (1993). Biotechnology and genetic engineering: What Europeans think about it in 1993. *Eurobarometer* 39.

INRA, Institut National de la Recherche Agronomique (2000). The Europeans and biotechnology. *Eurobarometer* 52.

Jasanoff, S. (2005). *Designs on Nature: Science and Democracy in Europe and the United States*. Princeton University Press.

Joss, S. and Durant, J. (1995). *Public Participation in Science: The Role of Consensus Conferences in Europe*. Science Museum.

Macnaghten, P., Kearnes, M., and Wynne, B. (2005). Nanotechnology, governance and public deliberation: What role for the social sciences? *Science Communication* **27**, 268–287.

Macrory, R. (1997). National Biotechnology Conference: Report of the Rapporteur. DETR.

Mayer, S. (2002). From genetic modification to nanotechnology: the dangers of "sound science." In: Gilland, T., ed. *Science: Can We Trust the Experts?* Hodder and Stoughton, pp. 1–15.

Mayer, S. and Stirling, A. (2002). Finding a precautionary approach to technological developments—lessons for the evaluation of GM crops. *J Agric Environ Ethics* **15**, 57–71.

Mehta, M. D. (2004). From biotechnology to nanotechnology: what can we learn from earlier technologies? *Bull Sci Technol Soc* **24**, 34–39.

MORI (1999). The Public Consultation on Developments in the Biosciences. Department of Trade and Industry.

Nordmann, A. (2004). Molecular disjunctions: staking claims at the nanoscale. In: Baird, D., Nordmann, A., and Schummer, J., eds. *Discovering the Nanoscale*. IOS Press.

Ravetz, J. (2005). The post-normal science of safety. In: Leach, M., Scoones, I., and Wynne, B., eds. *Science and Citizens: Globalization and the Challenge of Engagement*. Zed Books, pp. 43–53.

Sandler, R. and Kay, W. D. (2006). The GMO-nanotech (dis)analogy? *Bull Sci Technol Soc* **26**, 57–62.

Tait, J. and Levidow, L. (1992). Proactive and reactive approaches to regulation: the case of biotechnology. *Futures* **24**, 219–231.

Waddington, C. H. (1978). *The Man-Made Future*. Croom Helm.

Wilsdon, J. and Willis, R. (2004). *See-through Science: Why Public Engagement Needs to Move Upstream*. Demos.

Wolfson, J. R. (2003). Social and ethical issues in nanotechnology: lessons from biotechnology and other high technologies. *Biotechnol Law Rep* **22**, 376–396.

Wynne, B. (1992). *Rationality and Ritual: the Windscale Inquiry and Nuclear Decisions in Britain*. British Society for the History of Science.

Wynne, B. (1995). Public understanding of science. In: Jasanoff, S., ed. *Handbook of Science and Technology Studies*. Sage, pp. 361–388.

Wynne, B. (2001). Creating public alienation: expert cultures of risk and ethics on GMOs. *Science as Culture* **10**, 445–481.

Nano and Bio: How are they Alike? How are they Different?

Paul B. Thompson

The creation of a Social and Ethical Implications of Nanotechnology (SEIN) component in the National Nanotechnology Initiative (NNI) was a response to two observations on the part of United States science and legislative leadership. One was the recognition that the Ethical, Legal and Social Issues (ELSI) program launched in connection with the Human Genome Initiative at the National Institutes

What Can Nanotechnology Learn from Biotechnology?
ISBN: 978-012-373990-2

of Health (NIH) and the Department of Energy (DOE) had been successful in identifying and articulating significant legal, ethical, and policy issues associated with the sequencing of the human genome and with the rise of genomics, in general. The second was the hope that a similar effort might have been able to either forestall or at least prepare the science community for the kind of public reaction that had been experienced in connection with agrifood biotechnology and so-called genetically modified organisms (GMOs) (Berube, 2006).

Indeed, the claim that nanotechnology must avoid the mistakes and pitfalls of "the GMO debacle" is so frequently expressed in connection with this emerging area of science and technology as to be virtually ubiquitous. This warning is repeated in studies by the US National Research Council (NRC, 2002) and the British Royal Society (Royal Society and Royal Academy of Engineers, 2004). David Berube's book *Nano-Hype* quotes numerous scientists, business leaders, and public officials who express this sentiment, including Mike Rocco, Rita Colwell, and Sean Murdock of the Nanotech Business Alliance. His index lists 28 separate references to GMO foods, several of them extending over several pages (Berube, 2006).

Crucially, the link between nanotechnology and GMOs is also cited frequently by non-governmental organizations (NGOs) critical of nanotechnology, including the ETC Group (2004) and Greenpeace (Parr, 2003). It is with these forebodings in view that other chapters in this book have distilled lessons from the experience with agrifood biotechnology that budding nanotechnologists might bear in mind.

But what if all these voices are wrong? What if agrifood biotechnology and the diverse range of nanotechnologies currently being developed are so different that there are no lessons to be learned? And if there are any lessons one of them must certainly be that public fear and outrage is quite selective, emphasizing specific applications of biotechnology at specific times and locations while leaving others wholly unscathed. As such, the goal of this chapter is to turn something of a skeptical eye on the premise of entire volume, and to examine some reasons for thinking that agrifood biotechnology and nanotechnology are more different than alike, at least as far as the potential for debacle is concerned. As might be expected, the hypothesis that the GMO experience has no relevance to nanotechnology is rejected, but the "high relevance" hypothesis is also sharply qualified. In this exercise, some of the most critical lessons from the debate over agrifood biotechnology and GMOs become more sharply distinguished.

Why nanotechnology may not be much like biotechnology

Part of the appeal for using debate and outrage over agrifood biotechnology to motivate SEIN research on nanotechnology lies in the obviousness of an association between emerging technologies that have indisputable similarities. Among these similarities are the promise of new and transformative products, an uncertain regulatory framework and the fact that non-scientists grasp an initial and intuitive understanding of what is work in each field that fades quickly into technical complexity and utter opacity with respect to the scientific details. But these are rather vague similarities, so any attempt to subject the hypothesis that nanotechnology is in for a bumpy ride in the domain of public perception must move to some more specific points on which a point-by-point comparison can be made. Ronald Sandler and W. D. Kay have made an initial attempt to do this in a paper that, like this chapter, questions the suggestion that there are any useful analogies between GMOs and nanotechnology.

Sandler and Kay make two important claims in their paper. First, they present an argument to show that nanotechnology is unlikely to experience the same kind of public controversy associated with GMOs. This is the argument that will occupy us later. Second, they sketch a series of possible lessons for scientists and administrators working in nanoscience and nanotechnology, arguing that most commentators who draw upon the analogy between GMOs and nanotechnology focus on the relatively shallow lesson of avoiding a negative public response. In contrast, deeper lessons would stress the need to involve the public in decision-making for nanotechnology, minimally through public participation in risk management and the establishment of research priorities, but ideally through the creation of open forums that would actively seek to define broad objectives and constraints for nano-scale R&D. Their main point is that shallow analogies with biotechnology make it less likely that scientists and engineers will engage the public in a manner that provides some opportunity for nanotechnology's future to be determined by a cross-section of society that includes more than technical experts, government science programs and venture capitalists. For this reason, they regard the analogy to biotechnology as unfortunate and unproductive (Sandler and Kay, 2006).

Here, Sandler and Kay's reasons for thinking that nanotechnology will not provoke the same kind of public and interest group response will be the focus, but it is important to recognize and endorse the larger point they make in support of extended, deep, and non-strategic public engagement. This chapter will conclude with a similar recommendation, though the skepticism with which Sandler and Kay regard the analogy between agrifood biotechnology and emerging nanotechnologies will also be somewhat qualified. In short, they identify five points on which agrifood biotechnology and nanotechnology differ:

1. Food technologies are particularly sensitive because they are ingested directly into our bodies.
2. Foods are culturally associated with "naturalness" while many non-food technologies are already regarded as "artificial."
3. The "playing God" argument pertains only to the alteration of living organisms.
4. While biotechnologies are designed to be released into the environment, nanotechnologies are designed to be contained.
5. The early emphasis on public education and on addressing social and ethical issues creates at least the appearance of responsible development, while this was not the case for agrifood biotechnology.

Each of these points rests on a plausible hypothesis to explain why agrifood biotechnology was received with suspicion and public resistance. If nanotechnology is really different from biotechnology with respect to the key elements in each hypothesis, then there is no reason to predict that nanotechnology will be received by the public in a similar way. However, all five of these hypotheses are nested within a broader set of assumptions about technological controversy that stress the role of effective communication between the developers of technology and the broader public. Biotechnology is thought to have fallen on hard times because these five areas of sensitivity were not addressed through processes of disclosure, discussion, consensus building, and risk communication.

But Sandler and Kay's analysis neglects some of the hypotheses for explaining resistance to biotechnology that are most frequently cited by participants in and analysts of the controversy. Thus in addition to these five ways in which to consider whether biotechnology and nanotechnology are more different than alike, we can add five more:

6. Agrifood biotechnology became defined in terms of two specific techniques, genetic engineering and animal cloning, while nanotechnology is a diverse set of techniques and research activities defined by the scale at which mechanisms operate.
7. Agrifood biotechnology was presented to the public in the form of specific applications—herbicide-tolerant and pest-protected crops—that provided no benefit to food consumers.
8. Lack of public confidence in European regulatory systems accounted for the differential degrees of public acceptance for GMOs in Europe and in the United States. In particular, "process not product" and "substantial equivalence" were a bust.
9. Agrifood biotechnology precipitated an extensive expansion of intellectual property rights both with respect to the types of processes and entities for which property rights are claimed (e.g. genes and living organisms), with respect to the types of individuals and organizations active in pursuing property claims (e.g. public sector scientists and universities) and with respect to scope of property claims (e.g. TRIPS and GURTs).
10. Agrifood technology came along at a time when shifts were occurring in the balance of power among supply chain actors.

We can gain some insight into the lessons that nanotechnology can take from biotechnology by considering in some detail how nanotechnology and biotechnology compare with respect to each of these ten points. In particular, it is worth asking whether it is plausible to think that an emphasis on more democratic participation in decision-making, risk management, and agenda setting would be an effective response to the hypothesized triggers for public outrage noted in each of these ten hypotheses.

As such, the analysis of ten hypotheses here focuses primarily on the way that claims about nanotechnology or biotechnology are understood to provoke negotiable areas of risk or concern, areas where better communication or more effective discourse in the public sphere might be expected to make a difference. This is especially the case for the five hypotheses proposed by Sandler and Kay, where the assumption that education and participation were lacking in the agrifood case has been placed in the foreground. There may be other ways that nanotechnology is "like" or "different from" biotechnology that revolve around issues of entrenched social or economic power associated with key actors such as governments, corporations, or universities, or that speak to more widespread social forces that generate movements of social protest and resistance. Some of the last five hypotheses speak to these alternative possibilities.

Hypothesis 1: food technologies are sensitive

The suggestion that food technologies are especially sensitive is given some credence by the fact that while GMOs and proposals for cloning livestock have met with outrage, revulsion, and opposition, the use of genetic engineering to develop new drugs and medical therapies has not met with a similar response (Midden *et al.*, 2002). But this fact also undercuts Sandler and Kay's putative reason for thinking that food technologies are particularly sensitive, for like foods, drugs and medical procedures are ingested directly into our bodies. Risks associated with drugs and medical therapies are typically thought to be voluntary and accepted under conditions of informed consent, though bioethicists have argued that these norms are frequently ignored by physicians and biomedical researchers. Perhaps it is more plausible to think that medical uses of biotechnology are more acceptable to the people who use them because their condition of medical distress makes them more likely to view the risks associated with medical interventions of any kind—including genetically engineered drugs or therapies—favorably. They are more socially acceptable because we are loath to deny others the opportunity to seek treatment and accept the risks of treatment.

Although these points of contrast between medical and food biotechnology help flesh out the sense in which food technologies are sensitive and thus prone to the kinds of reaction that greeted agrifood biotechnology, one should not forget that Sandler and Kay's claim might well strike the average food or agricultural scientist, not to mention the typical food-processing firm, as utterly mystifying. Foods and food production were subjected to an extraordinary degree of technological transformation throughout the twentieth century, including the development of novel ingredients and preparation methods, the rise of chemical fertilizers and pesticides, and the development of synthetic flavoring agents and preservatives. Consumers as a group embraced not only a series of unfamiliar cuisines and fast or frozen delivery systems, but also wholly novel foods such as oleomargarine, non-dairy dessert toppings and coffee creamers, and the famous frozen lemon cream pie that contained neither lemon nor cream. What is more, these developments were accompanied and sometimes made possible by a steady growth in the methods available for introducing and manipulating genetic novelty in plants, including several non-transgenic techniques that make it possible to cross species lines or to stimulate wholly novel mutations. Consumers accommodated themselves to broccoflower, diploid strawberries, and tomatoes capable of withstanding a 30 mile an hour impact when hurled from automated harvesting machines into the back of a truck. Given this background, how could anyone think that food technologies are particularly sensitive?

The point, of course, is that sensitivity is an interpretive construct that may reflect a perspective or point of view far more than it reflects any robust feature of the products and practices in which either nanotechnology or biotechnology become applied. It is also worth pointing out that there will be food nanotechnologies. Food packaging and processing applications of nanotechnology are already in use and more will certainly arrive in the first wave of products. So even if Sandler and Kay are right about food technology being especially sensitive, we must note that nanotechnology is a food technology. So rather than interpreting hypothesis 1 as a reason to think biotechnology and nanotechnology are different, perhaps this is actually a way in which they are alike. There are food and non-food applications in both domains. Thus one question of interest is whether the controversy over agricultural biotechnology had any spillover effect on other domains of biotechnology. The answer to this question is that while medical biotechnology appears immune to such effects, perhaps for the reasons noted above, there is no question that the controversy over food biotechnology has had a chilling effect on the use of genetic engineering for forestry, for wildlife conservation and for pollution abatement.

Hypothesis 2: the naturalness thing

As an adjunct to the sensitivity of foods, Sandler and Kay note that foods are thought of as "natural" while many other areas of technology are thought of as artificial. Thus, if nanotechnology comes to be thought of as an artificial means in the manner that technology is typically thought of as artificial, then it will not create dissonance with a preconception people typically associate with food viz. that it is natural. Many of the qualifications already noted in connection with hypothesis 1 apply here as well, but the belief that biotechnology is "unnatural" has certainly surfaced frequently in both lay and expert debates about agrifood biotechnology. As Mark Sagoff notes in his widely read article "Genetic engineering and the concept of the natural," the food industry has long promoted its products by claiming that they are natural and by professing loyalty to "Mother Nature." If there is an inconsistency between the food industry insider's belief that foods have been subjected to frequent technological innovations through the last century and the general public's belief that foods are nature's bounty wholly unsullied by technological artifice, the food industry itself is largely to blame for this situation (Sagoff, 2001).

In looking at the debate over agrifood biotechnology, concerns about "naturalness" are among its most ambiguous and puzzling points of contention. At least two distinguishable lines of thought can be discerned. One, prominent especially in European concerns that biotechnology is "against nature," stresses the view that foods and food production should be finely tuned and adjusted to local ecologies, and ecology is here understood to include the history and culture of human adaptation to specific places. Thus, the French concept of *terroir* reflects an aesthetic and cultural attachment to the relationship between soils, microclimates, and traditional farming or processing methods. To take an action that is "against nature" in this sense is to violate or ignore the accumulated wisdom of tradition in farming, brewing, cheesemaking, and the like. These traditional practices are thought to respect the integrity of the local environment as well as that of the materials with which they work. This notion would apply equally to farming and food as to other craft-oriented activities such as woodcarving, stone masonry, or the weaving of thatch (Thompson, 2003).

These practical understandings of what is natural may overlap with philosophical, literary, and religious interpretations of nature, many of which had their heyday in the Romantic period of the mid-nineteenth century, when the rising tide of industrialization threatened craft industries, village economies, and the mentality of intense local solidarity. Philosophers schooled in medical bioethics associate this mentality with racial and ethnic prejudices and with a rhetoric in which claims that a given practice is "unnatural" means that it is foreign and contrary to the reigning social order. Thus they are familiar with arguments alleging that it is unnatural for women to use birth control or to work outside the home, much less to become doctors and scientists. They associate the concern for naturalness with eugenics and with socially conservative views about sexuality and family life, and they have brought an arsenal of arguments to combat these views into the arena of debate over biotechnology. And not all bioethicists oppose such socially conservative viewpoints. Thus, some have found cloning of livestock and genetic engineering of food "repugnant" and "monstrous" in just the way that they find homosexuality or abortion on demand repugnant and monstrous (Kass, 1997; Midgely, 2000).

How is nanotechnology alike and different from biotechnology with respect to hypothesis 2? Sandler and Kay may be correct in thinking that nanotechnology will escape concerns about what is natural and what is not, but a more careful review of how this concern has played out in the debate over GMOs suggests that we should simply wait and see. The deeper analogy to agrifood biotechnology here may draw less on what is meant by saying that a technology is or is not natural than on the way that a set of philosophical and political perspectives formulated within bioethics were applied in a somewhat thoughtless and ill-informed manner to a different domain of technological application. Medical schools and biomedical research funding agencies have supported teaching and research on social and ethical topics in their own areas of specialization far more aggressively than other areas of science and technology. The social scientists, philosophers, and historians who work on medical issues have well-developed networks among themselves and ties to the international media. This may well mean that medical nanotechnology will have excellent research capability for social and ethical dimensions, but that other domains of nanotechnology will suffer from a quick and dirty re-application of findings from the medical arena, with little serious attention to the circumstances, issues and constituencies that are truly relevant beyond the medical arena.

Hypothesis 3: playing God

Sandler and Kay argue that the "playing God" argument applies to living organisms, but not to the non-living processes that are most widely associated with nanotechnology. Again, many of the points already discussed could be made anew with respect to this hypothesis. In particular, the "playing God" concern has been prominent with debates over the possibility of applying to genetic engineering to human beings. A few prominent critics of agrifood biotechnology, notably Prince Charles in the UK, have used this kind of language in opposing GMOs, but to a significant extent this is a "concern" about biotechnology that was imported into the debate over farming and food, rather than arising within it. Again, the lesson, which seems deeply relevant to any kind of technology that is not well understood by the general public, is that even well-educated and highly trained people are likely to be fairly sloppy in the way that they cross-fertilize and apply critiques from one domain of technical application to another.

A close look at the "playing God" concerns as they have actually been articulated in debates over biotechnology reveals three themes. One is a straightforward concern about technological advances of all kinds, one that in Christian religious symbolism becomes associated with the fall from grace and by extension with human hubris and avarice. Clearly, manipulation of living organisms has nothing to do with this kind of concern and there is no reason to think that nanotechnology will be immune from it. A second concern, again more closely tied to human biotechnology than to food, is the possibility that a given application of the technology might come in conflict with the religious rules and rites of a particular group, with their theological doctrines or with their general religious sensibility. The hottest debates involve medical biotechnologies for human enhancement, and nanotechnology will step right into this controversy without skipping a beat. This is not, of course, a lesson from the agrifood debate. Yet agricultural scientists and food industry firms have seemed without the will or resources to confront those questions about acceptable dietary practice that are raised by GM food, seemingly taking a "don't ask, don't tell," stance toward the possibility that the technology might require any kind of religious consideration. What is more, there was a Judeo-Christian bias to the literature that existed for a very long time, as if responsibilities had been met if the technology was cleared by the Pope, a few rabbis, and the occasional Protestant church council (Brunk and Coward, 2008).

The third sense in which "playing God" has arisen as a concern in the agrifood biotechnology debate was nicely analyzed by Allen Verhey (1995). Focused primarily (again) on medical biotechnology, Verhey argues that the main point of raising questions about playing God is not to claim outright that a specific use of technology violates religious practice, but rather to call for a contemplative, spiritual, and religiously informed consideration of the technology and its social context. On this view, biotechnology might be religiously problematic not because it involves the modification of living organisms, but because it arose and was disseminated in a social environment dominated wholly by profit-making sensibilities, or because key decisions were taken in forums where people of faith were excluded, or where specifically religious or ethical language and considerations were expunged from the vocabulary. Again, there is no reason to think that nanotechnology is any less likely to be subjected to this kind of concern than is biotechnology.

Hypothesis 4: environmental release

Sandler and Kay suggest that the public feared GMOs because they would be released into the environment, and because, as living organisms, they would spread on their own. To the extent that nanotechnology is, as they write, environmentally contained, this fear will not arise. There seems to be little basis on which one could object to this point, as far as it goes. Those nanotechnologies that can be environmentally contained are indeed unlikely to provoke resistance on environmental grounds. Yet on first blush this also seems to be a supposed difference that also turns out to be a similarity upon even cursory inspection. On the one hand, environmentally contained applications of agrifood biotechnology such as recombinant rennet do not seem to stimulate much active opposition. On the other hand, the environmental fate of nanoparticles has already emerged as a central concern for nanotechnology (Colvin, 2003) and a principal focus of activism for environmental NGOs.

Yet Sandler and Kay lay emphasis on a specific notion of the release and containment relationship. It is the transgenic organism's ability to reproduce and the possibility that transgenes will spread to wild relatives (which will then reproduce) that leads to a special fear among biotechnology's opponents about the risks of environmental release (see, for example, Rissler and Mellon, 1996). This is arguably distinct from ordinary pollution concerns in that the challenges of clean-up or mitigation for self-reproducing organisms are at least more daunting, if they can be met at all. But while this more focused statement of the hypothesis seems credible, it points toward three additional areas of complication. One substantive environmental issue is that the possibility for nanoparticles to interact with mechanisms of cellular reproduction cannot be excluded. As such, ecological impacts quite like those of transgenic organisms are within the realm of theoretical possibility (Colvin, 2003).

The second complication involves perception and interpretation of environmental release as much as they involve substantive environmental concern. The more fantastic scenarios for nanotechnology involving replicators and "grey goo" are every bit as frightening as GMOs with respect to environmental release, if not more so. Thus even if the nanoscience community comes to believe that environmental containment is not a serious issue for more realistic nanotechnologies, these images are available to activists and to the public at large. They allow others to conceptualize risks of nanotechnology in precisely the sort of runaway train scenarios that are alleged to have created problems for agrifood biotechnology. Many in the science community would allege that these scenarios are not credible, just as many have argued that the scenarios of replicators and goo have nothing to do with real-world nanotechnology. Again, the similarities seem more profound than any points of difference.

In sum, the question of environmental release does not seem to be a reason for discounting the analogy between GMOs and nanotechnology. Nanotechnology and biotechnology seem to be almost alike in this respect, with some applications raising questions about containment and environmental risk, and other applications being largely exempt from those questions. To the extent that biotechnology raised superficial and uninformed fears about technology running amok through the environment, there is every reason to think that nanotechnology can raise similar superficial and uninformed fears. The larger lesson is that it is a constellation of fears, grounded and ungrounded, along with serendipitous events and orchestrated campaigns that coalesce to form the basis for a social movement in opposition to a given technology or to a subset of applications associated with a technology. Environmental risk and containment issues are clearly among the constellation of concerns that contributed to public opposition to GMOs. But just as no single issue could have been predicted to be the single triggering event for public resistance, there is no way to argue that the presence of an environmental dimension would set off opposition to nanotechnology any more than its absence would make nanotechnology immune to such reactions.

Hypothesis 5: public educational efforts are inoculating nanotechnology against public opposition

Sandler and Kay argue that because nanotechnology has already been accompanied by organized and well-funded efforts to study social and ethical dimensions, and because efforts to educate the public about it are already underway, it is a wholly different case than agrifood biotechnology. In one sense they are saying that because developers of nanotechnology have learned the lessons of early public involvement from the experience of agrifood biotechnology, the experience of public opposition will not be repeated. As with some of their other points, the experience with medical biotechnologies lends some support to this hypothesis. The ELSI initiative of the Human Genome Project dedicated millions of dollars in NIH funding to research on the medical applications and social issues that might conceivably arise in connection with human genetics research in general and with the sequencing of the human genome, in particular. The fact that these potentially explosive social issues have not emerged as a source of opposition to either genomics or to medical biotechnology provides some level of support for thinking that undertaking the kinds of social, ethical and legal research that are now going on in the US with the support of the National Nanotechnology Initiative will do some good.

Yet there were efforts to educate the public about agrifood biotechnology, notably by agricultural extension service programs, and there was an extensive effort to hold a public conversation about agrifood biotechnology conducted by the Keystone Center through 1992. The Center published no less than seven documents summarizing these now little known attempts to address issues of public participation and public concern associated with agrifood biotechnology (Keystone Center, 1988a, 1988b, 1988c, 1989a, 1989b, 1990, 1992). Also in the late 1980s, the National Agricultural Biotechnology Council (NABC) was formed, with membership from leading non-profit research organizations, mostly consisting of land-grant universities. The NABC has conducted annual public forums on biotechnology since 1989, and has issued over a dozen reports. The first of these was focused on sustainable agriculture (McDonald, 1989), and subsequent topics covered animal transformation, safety regulation, and, in recent years, the turn from food-oriented biotechnology to pharmaceutical and industrial crops. The 2006 meeting focused on industry–university partnerships. While participation from the broad public and opposition groups has waned over the years, the NABC conferences have arguably done more to educate research administrators in the public sector about social and ethical implications of agrifood biotechnology than they ever did to educate the public.

So biotechnology and nanotechnology are similar in this respect at least: leaders in both fields believed that they were undertaking responsible efforts to communicate with broader public and key constituencies, including environmental groups. The fact that hardly anyone remembers or refers to the Keystone effort a little more than 10 years after its completion, or is aware that ongoing NABC events continue should give pause to those who think that public education efforts produce a kind of resistance to resistance. So without undercutting the main force of hypothesis 5, which is to suggest that making an effort to inform and involve people beyond the technical fields might actually do some good, the concluding theme is that again nanotechnology may be very much like biotechnology precisely with respect to public education, despite the fact that the nanotechnology community has been congratulating itself as being different.

Hypothesis 6: agrifood biotechnology was narrow, nanotechnology is broad

Sandler and Kay's five hypotheses are framed under the presumption that consumer and activist reactions to GMOs were failures in communication. Either the developers of these technologies failed to adequately educate the public in the scientific principles and empirical findings that supported the development of agrifood biotechnology, or they failed to hear and understand a set of legitimate concerns that were being voiced by critics. Of course, it is entirely possible that both of these things were true. Now we turn to a new set of hypotheses that stipulate ways in which biotechnology and nanotechnology might be different not so much in the communication challenges they present as in the technical, governmental, and marketing hurdles that must be cleared. These hypotheses presume various ways in which opposition to GMOs might have been the result either of broad socio-culture trends that would have been difficult (if not impossible) to address simply through better communication, or of strategic failures in product development, marketing, or regulation. In either case, the lessons for other emerging technologies might not have very much to do with communication or public engagement.

The suggestion behind hypothesis 6 is that opposition to agrifood biotechnology converged on two specific technologies: mammalian adult–cell cloning and genetically engineered crops. Nanotechnology, in contrast, is a million different things that are already being done and that might in the future be done at the scale of 100 nm or less. The idea is that public concern and opposition congeals around a few well-defined applications and without this point of focus it will be difficult for opponents of nanotechnology to get much traction with the wider public. In fact, transgenic plants became defined largely in terms of two main real applications, herbicide-tolerant and pest-protected crops, and two speculative ones, vitamin A fortified (or "golden") rice and the "terminator" gene construct for rendering seeds sterile. While there is serious research intended to realize these speculative applications, neither has moved beyond proof of concept and it is questionable as to whether either will ever be found in a commercially produced plant variety. In addition, the broad image of agrifood biotechnology associates the term with adult–cell mammalian cloning (the famous case of Dolly the sheep), a technique which does not involve gene transfer (Preist, 2001; Wagner *et al.*, 2002). Arguably these five emblematic applications gave those who wanted to raise questions about biotechnology and the international agrifood research complex a focus that could be communicated succinctly.

No such focus appears on the horizon for nanotechnology. All of the five emblematic biotechnologies were developed within public or private labs at least nominally committed to agricultural science and all five involve cellular manipulation. Four of the five involved interspecies genetic engineering. There is nothing remotely like that kind of coherence on the science, engineering, or product front for the possible applications of nanotechnology. If having a coherent story to tell about an emerging technology is important to the mobilization of publics who wish to become involved in its management and development (Toumey, 2004), this difference between agrifood biotechnology and nanotechnology may be extremely important. It may mean that there will be no way for the mass media to present a unified story line about nanotechnology. Furthermore, it may mean that there will be very little coherence or overlap in the specific social, environmental, or economic interests that have the potential to mobilize people to express concern about nanotechnology. The upshot is that alleged lessons about the need to have better education and public involvement for nanotechnology are just a lot of hot air.

But as before, it is important to give this hypothesis a critical examination. At the early stages, the same sort of incoherence observation might have been made about agrifood biotechnology. In the 1980s, people in agriculture and the food industry were using the term biotechnology to include everything from tissue culture to ordinary brewing (Busch *et al.*, 1991). The point here is that what biotechnology is taken to be by the broader public is very much shaped by public discourse, including and especially media coverage. Media coverage in turn is shaped by what sources say to reporters when they are interviewed (Priest, 2001). In the case of biotechnology, there was an early tendency to see many distinct techniques and products as examples, in part because association with the biotechnology trend was thought to be a good way to attract funding of all kinds. Organizations such as the International Food Information Council conducted numerous studies on what various terms meant to members of the public, and which ones would elicit the warmest responses. However, the term GMO—favored by absolutely no one in the science community—is the one that stuck, and the term "biotechnology" came to be associated not only with genetically engineered crops and foods, but also with mammalian cloning (Wagner *et al.*, 2002).

Thus, the association of GMOs and biotechnology with a narrow set of genetically engineered crops and cloning is actually a bit of an overstatement, at least as far as the underlying science is concerned. There is less coherence here than might originally be thought, and it is entirely possible that 20 years from now the term "nanotechnology" will be understood to mean a relatively restricted set of tools, techniques, and applications. One cannot deny that at present nanotechnology—spanning information processing, nanolithography, microscopy, drug delivery, and materials science, to name only a few—does not exhibit much coherence, so one must admit that the difficulty of getting a handle on what nanotechnology actually is may in fact slow down public opposition. Frankly, it is simply too early to tell whether nanotechnology has the potential to become strongly associated with any specific and emblematic applications. At present, nanotechnology could be many different things, and in this respect it is different from what agrifood biotechnology is now typically thought to be. But if one were to consider nanotechnology and biotechnology at comparable points in their development, it is far less clear that they are different. The lesson that SEIN research can offer here is that specific products, even products that are never developed or commercially released, can create a lasting impression on the public mind. Scientists and the business community would be foolish to think that the "official" definition of nanotechnology is the one that will stick in the public mind.

Hypothesis 7: no benefit to consumers

One of the most frequently repeated "lessons" among agrifood biotechnology "insiders" (that is, the scientists, investors, and company or experiment station officers charged with public relations and management), is that the trouble arose because the benefits of the first-generation products were for farmers rather than food consumers. As such, consumers could see little reason to accept any level of risk associated with the product. Even speculative risks are too high when one is deriving no benefit in exchange for bearing them. Support is given to this hypothesis by the fact that recombinant rennet (that is, chymosin, the enzyme used in cheesemaking, produced by genetically engineered bacteria) caused absolutely no stir anywhere and is now used widely in countries that rejected genetically engineered crops. The reason proffered is that the recombinant product is both purer and, unlike natural rennet, is not harvested from slaughtered calves. The latter fact makes the recombinant version compatible with kosher dietary practice and is also attractive to animal rights advocates. As such, the product was thought to have benefits to cheese consumers (Ahson, 1997; Mehta and Gair, 2001). The degree of support for hypothesis 7 that can be derived from the recombinant rennet example is mitigated by the fact that very few consumers know that recombinant rennet has been widely adopted by the cheese industry.

Hypothesis 6 appealed to fairly broad generalizations about the mobilization of interest groups and broader publics to suggest that it would be difficult for resistance to form in response to the relatively diffuse cloud of techniques and technologies coming forward under the banner of nanotechnology. Hypothesis 7 suggests that fairly specific strategic errors made by for-profit firms led to the problems with agrifood biotechnology. It too rests upon some broad generalizations about human behavior, to wit, that people tend to make decisions that accord with their perceived self-interest. The theory behind the "no benefits" hypothesis involves an application of this generalization to decision-making involving potential risks. Consumers might be willing to try something new (i.e. take a risk) if they had some reason to think that it was beneficial to them, but since there were no product-attribute benefits associated with GMOs, why should they try them?

As an analogy to nanotechnology, the response here must be very much like the response to hypothesis 6: It is simply too early to tell whether the products most prominently associated with nanotechnology will have benefits to consumers. Early non-food biotechnologies had little impact on the public's receptiveness to GMOs, so public discussions on nanotech medical devices prove nothing either way. The key test will come when some product of nanotechnology becomes associated with a public health or environmental risk, with human exposures occurring through air, water, or food. If the rationale for that product resides in efficiencies for upstream producers in the supply chain, rather than for endpoint consumers, then nanotechnology will be very much like biotechnology, at least if we set the example of recombinant rennet to the side. As a potential lesson for product developers in nanotechnology, the point is to be sure that the first generation of products that can be conceived as risky are products that consumers will perceive as very attractive and beneficial to them.

More broadly, hypothesis 6 is one example of many possible business decision-making mistakes that technology companies can make. The literature on agrifood biotechnology and GMOs is replete with analyses articulating specific mistakes made by key individual companies or actors. Dan Charles' book, *Lords of the Harvest*, provides a highly readable overview of several additional mistakes that supplement the "no benefits" argument examined above. Charles discusses the Calgene Company's Flav*rSav*r tomato, expected to be one of the first food GMOs to reach the market and thus widely regarded as an important test case. The tomato was a commercial flop, but not because of genetic engineering. This product did provide taste and quality benefits that might been attractive to consumers, but Calgene made some elementary mistakes in selecting the right crop cultivar to use its antisense technology, and the tomato proved impossible to ship and market in leading grocery stores. Another notable failure described by Charles concerns the Monsanto strategy for moving its successful transgenic crop varieties into European markets. Here, Charles argues that Monsanto simply adopted an approach that was perceived as arrogant, brash, and hasty, providing an open opportunity for protectionist groups to promote anti-biotechnology sentiments that eventually congealed into entrenched consumer resistance (Charles, 2001).

David Sparling's chapter in this book (Chapter 9) provides a succinct source of the lessons that can be learned here, especially by scientists and engineers who become involved in start-up firms to develop new technology. For all manner of bad decision-making, the determination of whether agrifood biotechnology is or is not a good analogy to nanotechnology depends on what happens in the future. It depends on the decisions that developers of nanotechnology actually make. The SEIN researches that will be relevant to avoiding bad decisions are, again, not necessarily forms of research that emphasize public participation or democratic decision-making. More conventional kinds of management and marketing research will almost certainly be more relevant, while the relative ignorance that most SEIN participation researchers have about the business decision-making environment will probably make their contributions appear naïve, at best.

Hypothesis 8: lack of confidence in the regulatory system

Of the possible explanations for European resistance to biotechnology, a crisis in Europeans' confidence in governments' capabilities for effective regulation is among those having a significant degree of empirical support from empirical research (Gaskell *et al.*, 2002). This research suggests that biotechnology might be a poor source of lessons for nanotechnology because nanotechnologies will face a somewhat different and presumably more stable regulatory environment. Thus, hypothesis 8 states that nanotechnology will be different from agrifood biotechnology in that the primary cause for European public resistance to biotechnology (the crisis of confidence in the European regulatory system) will have been removed by the time that nanotechnologies begin to move through the regulatory system in significant numbers.

In evaluating this hypothesis, a great deal turns upon the alleged link between confidence in the regulatory system and the upsurge of public concern about GMOs. This link has been the focus of several public opinion studies (Gaskell *et al.*, 1997, 1998, 1999, 2000) but can be further grounded in several considerations that are more difficult to evaluate with surveys and other kinds of empirical study. For one, Europe was undergoing a period of political uncertainty as national regulatory systems were being harmonized under the provisions of the European Union. This put inconsistencies and disputes about regulatory standards into the headlines, and also meant that agrifood biotechnology entered European regulatory review at a time when regulators themselves were unsure how to proceed. As such, GMOs bore the brunt of internal procedural uncertainty within the regulatory community (Toumey, 2006). Another factor is the European experience with several high-profile regulatory failures. "Mad cow disease" was the most obvious, but there were other failures in food safety regulation across Europe that eroded public confidence. What is more, ongoing coverage of contamination from the Chernobyl accident made Europeans leery of regulatory decisions made elsewhere (Jasanoff, 1997).

Europe is spending heavily on research in nanotechnology (Berube, 2006, pp. 137–144), and one would expect that European regulators will be prepared to address regulatory issues associated with products from nanotechnology. As such, there is some reason to think that the lack of clarity in regulatory procedures that arose in connection with the process of harmonization will not continue to plague new technologies in Europe. However, it is much less clear how members of the European public will respond to nanotechnology irrespective of their confidence in regulatory agencies, and disagreement is possible (witness contributions to this volume from

George Gaskell, Chapter 12, and from Phil Macnaghten, Chapter 6). An interesting corollary to this hypothesis relates to the assumption that SEIN research and education efforts in nanoscience and nanotechnology will promote public acceptance. This assumption is based on research that demonstrates a correlation between "science literacy" and support of new technology in the United States. Europeans generally have much higher levels of literacy in science than do Americans (Evans and Durant, 1995), but this did not promote acceptance with respect to biotechnology. In fact, better understanding of science is believed to have made Europeans more skeptical and able to recognize failures in the regulatory process (Buchmann, 1995).

Yet it may be worth injecting a note of skepticism about the "regulatory confidence" assumption. On one hand, it has never been especially clear to me that the alleged American confidence in regulatory agencies is very robust. The way that hypothesis 8 is generally understood in the US context is that the "good science" used by US regulatory agencies is what leads to higher levels of public confidence in the regulatory process, but I demur. Here, an anecdote must suffice. I spent a long day at a conference in about 1999 listening to people touting Americans' confidence in the regulation of food and environmental biotechnology as a type of "vaccination" against the reactions then rampant in Europe, only to go back to my hotel room and hear US Senator Trent Lott (R—Mississippi), then Majority Leader, on national television saying "Well, of course, nobody believes the FDA." Lott was talking about drug approval rather than GMOs, but the point is that fissures and seams abound in every political culture, and the possibility that public confidence will disappear into a crevasse on a moment's notice cannot be dismissed.

On the other hand, the differences between the United States and Europe with respect to an average citizen's ability to exert influence over the direction of science and technology extend far beyond those relevant to that moment in time when Monsanto attempted to introduce GMOs in Europe. Chris Toumey has argued that "science and democracy" movements have a long history in the US that they lack in Europe (Toumey, 2006), and it is also worth noting that the differences between liability and administrative law provide US citizens enormous opportunities to influence both government and industry through the mechanism of the lawsuit. So European-inspired moves to address confidence through various participation mechanisms may yet pale in comparison to the power residing in the hands of any American who can afford an attorney. In sum, nano-technology has the potential to be different from agrifood biotech-nology with respect to public confidence in the regulatory process, but it remains to be seen whether it will differ in fact. There are a lot of opportunities for SEIN research to clarify and contribute to our understanding of the relationship between public confidence, public resistance, and the regulatory process.

Hypothesis 9: intellectual property rights

Critics of agrifood biotechnology had success in caricaturing the first generation of products as instances of overweening corporate greed, and a large part of that success must be attributed to the fact that companies were quite obviously and undeniably busy trying to stake a number of new ownership claims in connection with biotech. These included fairly straightforward process patents that made news on the business page, but also patents on whole organisms, patents on sequences of genetic code (with and without characteri-zation of function), licenses and technology fees that would be required of farmers who wished to plant transgenic crop varieties, and genetic use restriction technologies (GURTs), most prominently the so-called "Terminator" gene, that could be used to protect intellectual property through technical means (Priest, 2001). Biotechnology also became associated with several high-profile legal disputes, includ-ing eventually disallowed claims over properties associated with the Indian neem tree and Monsanto's actions against Canadian farmer Percy Schmeiser for theft (Ziff, 2005).

Why would one think that agrifood biotechnology would differ from nanotechnology with respect to intellectual property? It is cer-tain that patents will be sought for processes and products of nano-technology. The ETC Group Report *Down on the Farm* lists over 30 patent applications for nanotechnologies in the field of food packaging and food technology alone (ETC Group, 2004). But there are two reasons to think that patents in agrifood biotechnology have a special resonance with interest groups and the broader public. One arises in connection when the ethical force of concerns about alter-ing life forms is joined to claims of ownership. Many critical reviews of the intellectual property disputes in agrifood biotechnology were led by headlines or titles in which the phrase "owning life" was prominently displayed. On this front, even biomedical biotechnologies were not immune. The book *Who Owns Life?* provides a representative sample of both medical and food-related essays raising concerns about the use of various mechanisms to control the use of discoveries in genetics (Magnus *et al.*, 2002).

The second reason relates more narrowly to agriculture. Patents were seen as an important component in the three leading agricultural biotechnology companies' attempt to expand their control over a broad array of farm supply technologies, especially seeds and chemicals. Expansion here means both that there was an obvious reduction in the number of firms marketing these products to farmers in North America between 1980 and 2000, and that the companies which emerged from the competition as industry leaders were making aggressive attempts to expand into global farm supply markets (Kalaitzandonakes, 1998). This latter activity has been especially troubling to NGOs focused on poor, low-resource farmers in the developing world—groups, NGOs, that is, like ETC Group. What is more, previously existing forms of intellectual property protection in agriculture, such as existed under the US Plant Variety Protection Act (PVPA) of 1970 had preserved farmers' right to replant seed saved from a previous years' crop, though they were banned from commercial production. The transformations of intellectual property in agriculture were not simply extensions of intellectual property business as usual, especially from the perspective of working farmers.

Thus, it is plausible to think that patents and licenses for traits and processes in agrifood biotechnology had something of a unique effect in precipitating social conflict and public outrage. Yet in what is becoming a recurring theme, we must conclude that it is, again, too early to say whether nanotechnology will be like or different from biotechnology with respect to claims and disputes over intellectual property rights. It does not seem likely that rhetorical ploys stressing the ownership of life or farmers' rights will be readily available with respect to nanotechnology. However, more substantive disputes over the distinction between discovery and invention that occurred in connection with attempts to claim ownership in biotechnology would seem to be very likely to recur as nanotechnology firms attempt to establish claims over nano-scale processes. Furthermore, nanotechnology would not appear to be immune from a third reason why intellectual property became controversial in agrifood biotechnology, that being the entry of public institutions into active pursuit of patents for research discoveries.

The US Bayh-Dole Act of 1980 allowed universities to seek patents for work done with Federal research dollars. This piece of legislation was not uniquely targeted to biotechnology, but it took a number of years before universities and non-profit research institutes developed the internal management structure to make full use of this opportunity. This ramp-up time coincided with the 1980 US Supreme Court decision in Diamond vs. Chakrabarty allowing the patenting of living organisms and the Animal Patent Act of 1986. Thus, whether real or perceived, the public, non-profit science sector's establishment of intellectual property offices intended to capture some economic return on their research activities has been strongly connected to biotechnology. A cover article in the *Atlantic Monthly* entitled "The kept university," articulates some of the public concerns associated with these developments (Press and Washburn, 2000). So while nanotechnology is unlike biotechnology in some of the respects that made patents a problem, it is not at all unlike in other respects. Here, as before, SEIN research should prove fruitful.

Hypothesis 10: changing relations of economic power

From the perspective of institutional economics, the agrifood biotechnology controversy is all about the attempts of people and groups occupying downstream positions in the supply chain to exert power on upstream actors. Whereas food companies such as Gerber baby foods or the Frito-Lay chip company once typically dealt only with commodity suppliers and expected only that the raw commodities meet government safety standards, with the advent of biotechnology they notified suppliers that they must certify their goods as "GMO free." A similar action occurred between UK grocery chains and the suppliers of the store-branded foods. Here, downstream actors in the supply chain are demanding more control over the production process for the products that they purchase. Earlier biotechnology battles fought in the US when some small-scale dairymen and companies such as Ben and Jerry's Ice Cream attempted to resist recombinant bovine somatotropin (bST), could also be seen as a battle over economic power, though the downstream users lost that battle. And in one sense, food consumers themselves could be said to be attempting to exert greater control over upstream practices in the supply chain by demanding labeling of foods developed using agrifood biotechnology. This demand has not been successful in the US, and the debate over labels can be interpreted as a struggle in which agricultural input suppliers such as seed companies attempt to prevent their customers' customers from being able to dictate conditions upon them (Maltsberger and Kalaitzandonakes, 2000; Muth *et al.*, 2002).

With this hypothesis about what went wrong with respect to agrifood biotechnology, we reach a possible explanation residing in deeply rooted relationships between technology, economics, and social power. It is not an explanation that has been widely circulated among members of the scientific community or the general public. Changing power relations in the food supply-chain do seem to be very relevant to the events that have transpired around agrifood biotechnology. One would expect that a lot of new food technologies will become implicated in these battles during the coming decade. But it is quite possible that this is a phenomenon that will not be of great importance outside the food system, and it may simply be a pendulum effect occurring in response to many decades when economic power in the food system became concentrated in the hands of grain companies and input firms, both of whom have benefited from innovations tied to commodity grade production. There are many industries where downstream actors have long specified a host of traits unrelated to health and safety. As such, this may well be a point on which agrifood biotechnology really is different from many if not all non-food nanotechnologies.

Nevertheless, technological change can be implicated very deeply in shifting power relationships among affected various parties. This is, after all, perhaps the most enduring insight from the social and political writings of Karl Marx (1818–1883), and one that set the stage for a great deal of twentieth-century social science. Marx's association with communist political philosophy and the use of his name by several of the twentieth century's most repressive political regimes have provoked too many opinion leaders in the biological and physical sciences or engineering into a knee jerk reaction of rejecting any and all social explanations that bring a whiff of Marxism to the nostrils. The ongoing shift toward greater relative power among food retailers has, of course, been underway for some time, and is, in a broad sense, quite compatible with the Marxist hypothesis that concentrations of capital will lead shifts in economic power. Technology (along with political power) simply becomes a means. One would certainly expect that nanotechnology will both precipitate and become embroiled in similar power struggles.

So perhaps the greatest lesson to take from hypothesis 10 is that one should not be too literal in applying the analogy between biotechnology and nanotechnology. In fact, none of the hypotheses that have been offered to explain resistance to food biotechnology translate readily into necessary and sufficient conditions that could predict public resistance to new technology. The five hypotheses offered by Sandler and Kay are best interpreted as reasons or rationales that render the events that surrounded controversy over GMOs in intelligible terms, terms that allow us to understand how some person or group might have reacted with suspicion, revulsion, resistance, or distrust. The second group of hypotheses rely on bodies of theory and evidence from the social sciences that tend to emphasize structural and functional models of society or the economy, or of the rational individual and the profit-seeking firm.

Analysis

The ten hypotheses provide a broad (though not exhaustive) set of points on which to compare nanotechnology to agrifood biotechnology. All ten points of comparison state reasons why agrifood biotechnology became the subject of a widespread, indeed global, controversy, and then go on to propose respects in which nanotechnology might well be very different. All of the suppositions regarding difficulties in the development of agrifood biotechnology would require significant additional amplification and development before they could be understood as explaining the controversy over GMOs. Any social explanation of the controversy would draw upon theories and assumptions about individual and group rationality, organizational behavior, social movements, globalization, and the social construction of risk, not to mention general theoretical approaches to social and economic behavior in domains such as innovation, trade, and public choice. Thus the analysis here must be somewhat incomplete, and it should be stressed that none of these hypotheses can be put forward as indicating causal conditions for controversy and resistance. Nevertheless, the ten hypotheses highlight features of agrifood biotechnology and the controversy around it that provide useful and suggestive starting points for planning and managing nanotechnology R&D, as well as educational, regulatory, and public consensus projects intended to ensure the widest possible consideration of interests and concerns in governing nanotechnology. It is thus important to consider whether biotechnology and nanotechnology are really alike or different with respect to these ten points.

The five hypotheses noted by Sandler and Kay articulate reasons why someone might have an objection to agrifood biotechnology. They stress features—whether real or perceived—about agrifood biotechnology that might lead someone to find this technology distasteful, threatening, undesirable, or socially unacceptable, either in general or in the specific form that it was presented to the public in the 1990s. The second group of hypotheses describes characteristics of agrifood biotechnology as it emerged in the agrifood system between 1980 and 2000 that made this technology particularly vulnerable to social protest and to conflicts over economic power. These hypotheses for the most part presume a built-in tension or competitive circumstance that sets the interests of various firms within an industry at odds, or that places social interests such as labor and capital, industry and environment or respective national economies into opposition. These conflicting social forces vie for dominance in various marketplaces and through attempts to manipulate regulation, the terms of trade and their own (as well as their opponents') public image. All technologies must pass through this gauntlet, and the last five hypotheses note characteristics in the products, timing, or strategy of agrifood biotechnology that proved to be infelicitous.

Sandler and Kay's five hypotheses state a number of cultural and symbolic features of foods and living organisms that may have made agrifood biotechnologies more likely to trigger reactions of fear, distrust, revulsion, or general resistance than would be associated with many applications of nanotechnology. There is common sense to all these points of difference that should not be overlooked by those who assert that nanotechnology is highly likely to fall prey to the same kinds of resistance and controversy that were associated with agrifood biotechnology. However, a close examination of each hypotheses reveals that there may be more similarity between these two domains of technology than initial analysis suggests. As such, there are lessons to be learned with respect to how one should and should not develop and promote nanotechnology with respect to each of the five points on which Sandler and Kay assert that nanotechnology will be different. Indeed, Sandler and Kay's putative dismissal of the analogy between GMOs and nanotechnology is rhetorical in that they really want to make the larger point that the analogy "obscures the reasons for public engagement and SEI research, their proper focus and objectives" (Sandler and Kay, 2006, p. 61). Their goal is to promote better activities of engagement and participation, not to suggest that a complacent public will make such activities unnecessary.

For the most part, the hypotheses in the second group emphasize characteristics of agrifood biotechnology and its history for which there is no ready analogy with nanotechnology at present. But in each case, applications of nanotechnology and business or regulatory strategies for its public release could be forthcoming that would trigger very similar circumstances. As such, it is simply too early to tell whether nanotechnology is like agrifood biotechnology with respect to the key elements noted in the last five hypotheses, or different. Much will depend on choices that are made by the people who develop and promote nanotechnology over the coming years. Here, it would seem, there are clear lessons to be learned, and each of the five hypotheses actually sketches a fairly large domain where SEIN research might be well utilized by scientists, engineers, and key decision-makers for nanotechnology. It is critical to emphasize, however, that this kind of research will be fairly useless if it is not actually taken up and applied by those decision-makers. There is absolutely no reason to think that simply conducting such research, presenting it to SEIN colleagues, and publishing it in social science or humanities outlets will have any palliative or educational effect on the broader public. It is the nanoscience community itself that is the audience for this kind of work, not the non-technical citizen.

References

Ahson, K. (1997). What is actually happening in agro-food biotechnology? *Br Food J* **99**, 263–267.

Berube, D. M. (2006). *Nano-Hype: The Truth Behind the Nanotechnology Buzz*. Prometheus Books.

Brunk, C. and Coward, H., eds (2008). *Acceptable Genes*. SUNY Press.

Buchmann, M. (1995). The impact of resistance to biotechnology in Switzerland: a sociological view of the recent referendum.

In: Bauer, M., ed. *Resistance to New Technology: Nuclear Power, Information Technology and Biotechnology*. Cambridge University Press, pp. 207–224.

Busch, L., Lacy, W., Burkhardt, J., and Lacy, L. (1991). *Plants, Power and Profits: Social, Economic and Ethical Consequences of the New Biotechnologies*. Basil Blackwell.

Charles, D. (2001). *Lords of the Harvest: Biotech, Big Money, and the Future of Food*. Perseus Press.

Colvin, V. L. (2003). The potential environmental impact of engineered nanoparticles, *Nat Biotechnol* **21**, 1166–1170.

ETC Group (2004). *Down on the Farm: The Impact of Nano-scale Technologies on Food and Agriculture*. ETC Group.

Evans, G. and Durant, J. (1995). The relationship between knowledge and attitudes in the public understanding of science in Britain, *Public Understanding Sci* **4**, 57–74.

Gaskell, G. and the European Biotechnology and the Public Concerted Action Group (1997). Europe ambivalent on biotechnology. *Nature* **387**, 845–847.

Gaskell, G., Bauer, M., and Durant, J. (1998). Public perceptions of biotechnology in 1996: Eurobarometer 46.1. In: Durant, J., Bauer, M., and Gaskell, G., ed. *Biotechnology in the Public Sphere: a European Sourcebook*. Science Museum, London, pp. 189–214.

Gaskell, G., Bauer, M. W., Durant, J., and Allum, N. C. (1999). Worlds apart? The reception of genetically modified foods in Europe and the U.S. *Science* **285**, 384–387.

Gaskell, G., Allum, N., Bauer, M. *et al.* (2000). Biotechnology and the European Public. *Nat Biotechnol* **18**, 935–938.

Gaskell, G., Thompson, P. B., and Allum, N. (2002). Worlds apart? Public opinion in Europe and the USA. In: Bauer, M. W. and Gaskell, G., ed. *Biotechnology: The Making of a Global Controversy*. Cambridge University Press, pp. 351–375.

Jasanoff, S. (1997). Civilization and madness: The great BSE scare of 1996. *Public Understanding Sci* **6**, 221–232.

Kalaitzandonakes, N. (1998). Biotechnology and the restructuring of the agricultural supply chain. *AgBioForum* **1**, 40–42.

Kass, L. (1997). The wisdom of repugnance. *The New Republic* June 2, 17–26.

Keystone Center (1988a). Issues raised by biotechnology: A Keystone Biotechnology Discussion Paper. July 1988, Report 34. Keystone Center.

Keystone Center (1988b). Workshop Summary: First Regional Workshop on Biotechnology for Public Interest Leaders. Austin, Texas; Keystone Environmental, Citizen, State and Local Leadership Initiative. July 1988, Report 35, Keystone Center.

Keystone Center (1988c). Workshop Summary: First Regional Workshop on Biotechnology, Madison, Wisconsin; Keystone Environmental, Citizen, State and Local Leadership Initiative. March 1988, Report 36, Keystone Center.

Keystone Center (1989a). Keystone National Biotechnology Forum, Interim Summary Report, Public Participation and Education in Biotechnology: Summary Report and Recommendations. February 1989, Report 31, Keystone Center.

Keystone Center (1989b). Biotechnology Decisionmaking: Perspectives on the Objectives of Public Participation. September 1989, Report 28, Keystone Center.

Keystone Center (1990). Workshop Summary: Southeast Regional Workshop on Biotechnology, Atlanta, Georgia; Keystone Environmental, Citizen, State and Local Leadership Initiative. June 1990, Report 26, Keystone Center.

Keystone Center (1992). The Keystone Environmental, Citizen, State and Local Leadership Initiative for Biotechnology, Issue Report Series, Biotechnology Decisionmaking: Perspectives on Commercialization. July 1992, Report 24, Keystone Center.

McDonald, J. F. ed. (1989). *Biotechnology and Sustainability: Policy Issues*, National Agricultural Biotechnology Council (NABC).

Magnus, D., Caplan, A., and McGee, G., eds (2002). *Who Owns Life?* Prometheus Books.

Maltsberger, R. and Kalaitzandonakes, N. (2000). Direct and hidden costs in identity preserved supply chains. *AgBioForum* 3, 236–242.

Mehta, M. D. and Gair, J. J. (2001). Social, political, legal and ethical areas of inquiry in biotechnology and genetic engineering. *Technol Society* 23, 241–264.

Midden, C., Boy, D., Einseidel, E. *et al.* (2002). The structure of public perceptions. In: Bauer, M. W. and Gaskell, G., eds. *Biotechnology: The Making of a Global Controversy*. Cambridge University Press, pp. 203–223.

Midgley, M. (2000). Biotechnology and Monstrosity. *Hastings Center Report* 30(5), 7–15.

Muth, M. K., Mancini, D. and Viator, C. (2002). US food manufacturer assessment of and responses to bioengineered foods. *AgBioForum* 5, 90–100.

NRC, National Research Council (2002). *Small Wonders, Endless Frontiers: A Review of the National Nanotechnology Initiative.* National Academies Press.

Parr, D. (2003). Foreword. In: *Future Technologies, Today's Choices.* Greenpeace Environmental Trust.

Press, E. and Washburn, J. (2000). The kept university, *Atlantic Monthly* **285**(3), 39–42, 44–52, 54.

Priest, S. (2001). *A Grain of Truth: The Media, the Public and Biotechnology.* Rowman and Littlefield.

Rissler, J. and Mellon, M. (1996). *The Ecological Risks of Engineered Crops.* MIT Press.

Sagoff, M. (2001). Genetic engineering and the concept of the natural. *Philosophy & Public Policy Quarterly* **21**, 2–10.

Sandler, R. and Kay, W. D. (2006). The GMO-nanotech (dis)analogy? *Bull Sci Technol Soc* **26**, 57–62.

Thompson, P. B. (2003). Unnatural farming and the debate over genetic manipulation. In: Gehring, V. V., ed. *Genetic Prospects: Essays on Biotechnology, Ethics and Public Policy.* Rowman and Littlefield, pp. 27–40.

Toumey, C. (2004). Narratives for nanotechnology: Aniticipating public reaction to nanotechnology, *Techné* **8**, 88–116.

Toumey, C. (2006). National discourses on democratizing nanotechnology, *Quaderini* **61**, 81–101.

Verhey, A. (1995). "Playing God" and invoking a perspective. *J Med Philos* **20**, 347–364.

Wagner, W., Kronberger, N., Allum, N., de Cheveigné, S., Diego, C., Gaskell, G., Heinßen, M., Midden, C., Ødegaard, M., Öhman, S., Rizzo, B., Rusanen, T., and Stathopoulou, A. (2002). Pandora's genes: Images of genes and nature. In: Bauer, M. W. and Gaskel, G. eds. *Biotechnology: The Making of a Global Controversy.* Cambridge University Press, pp. 244–276.

Ziff, B. (2005). Travels with my plant: Monsanto vs. Schmeiser revisited. *University of Ottawa Law Technol J* **2**, 493–509.

Internet references

Royal Society and Royal Academy of Engineers (2004). Nanoscience and Nanotechnologies: Opportunities and Uncertainties. The Royal Society. http://www.nanotec.org.uk/finalReport.htm. Accessed June 22, 2006.

"It's Like Déjà-Vu, All Over Again": Anticipating Societal Responses to Nanotechnologies

8

Amy K. Wolfe and David J. Bjornstad

What Can Nanotechnology Learn from Biotechnology?
ISBN: 978-012-373990-2

Introduction[1]

This book rests upon the premise that at least some of the lessons learned from societal experiences with agricultural biotechnology are applicable to the emerging world of agricultural nanotechnologies. In this chapter, we take a step back from the particulars of what these lessons may be to examine the underlying bases for determining what is, or is not, a valid lesson. In short, the theme of this chapter is that we need to develop a trans-disciplinary science that allows us to analyze societal responses to new and evolving technologies. This trans-disciplinary science would seek to organize our understanding of societal responses within and across technologies more systematically and comprehensively than now is possible. Perhaps more importantly, it would begin to fill the enormous gap in our ability to anticipate societal responses to a range of technologies.

This call for a new societal response science aims to encourage a broader perspective than typically is taken and to grapple directly with fundamental questions whose answers still remain elusive, despite decades of study. It is not intended as a criticism of researchers and analysts who currently study issues associated with society–technology interactions, but rather to focus attention to analytical needs that can inform the shaping of the societal institutions through which regulatory measures and other means of social choice may be implemented (Keller, 2006, p. 6). A variety of studies, both here and abroad, have highlighted how nanotechnologies can present new challenges to these institutions.

Regulatory topics currently are a major thrust of social-nano research. For example, Shatkin and Barry (2006) describe how traditional approaches to risk analysis can be extended to deal with potential health and safety issues suggested by early scientific studies. By contrast, Mihail Roco (2005, p. 129), of the National Science Foundation, proposes starting with human needs and aspirations, and analyzing how nanotechnologies may reshape them over time. Working through the International Risk Governance Counsel, Roco and co-author Ortwin Renn (Renn and Roco, 2006) applied the risk governance methodology to develop a detailed plan for dealing with risk management for nanotechnology as it progresses through its developmental cycle. Meanwhile, the US Environmental Protection Agency instituted a research program in support of its responsibilities and issued a white paper describing its approach to nanotechnologies (Savage, 2006; US Environmental

Protection Agency, 2007). Other agencies of the US Government, notably the Occupational Safety and Health Administration, and the Food and Drug Administration, are also prepared to assume increasing nano-related responsibilities.

Despite this undercurrent of activity, members of the public are largely unaware of even the most basic elements of what has been described as a potential nano-revolution. In late 2006, a survey funded by the Woodrow Wilson Foundation reported that 42% of the public had not heard of nanotechnology and another 27% had heard "only a little" (Hart Research Associates, 2006). Clearly, many people may be in for a surprise.

However, the surprise may not be that the public at large is uninformed, but rather what societal responses will be as awareness increases and decisions must be made. Our overarching question is: How many more times will we be "surprised" by societal responses to technologies, their by-products, and consequences? To address this matter, we need to answer three other fundamental questions:

1. Why is the same technology sometimes accepted and sometimes rejected in apparently similar circumstances?
2. To what extent can we accurately anticipate societal responses and acceptability?
3. How can, or should, society make better-informed decisions?

How many more times will we be "surprised" by societal responses?

Yogi Berra's phrase, "it's like déjà-vu, all over again,"[2] epitomizes this question. Time and again, developers and promoters of a technology or solution seem surprised that their offering meets with resistance, rejection, or outrage. Obvious examples include nuclear power plants, incinerators, and the use of genetically modified organisms in agriculture. There also are a number of less obvious cases of technologies touted for their effectiveness or benefits that fail to find success or acceptance from a societal perspective. The diversity of technologies and issues on which we have focused, separately and in collaboration, are among these cases: environmental remediation strategies, energy-conserving technologies, and low-altitude military training flights.[3] Experiences with these technologies leads us to think that it should not be necessary for researchers to start anew, or nearly anew, in seeking to understand or anticipate societal responses to new technologies, their by-products, and wastes.

Why is the same technology sometimes accepted and sometimes rejected in apparently similar circumstances?

On the one hand, our sense of déjà-vu in thinking about societal responses to varied technologies leads us to posit that at least some of what seems so familiar can be formalized into generalized principles. This volume contributes to this process by asking explicitly what lessons learned from societal responses to agricultural biotechnology can be applied to agricultural nanotechnology. On the other hand, the laudable goal of identifying lessons learned and applying those lessons in a new realm is countered by the fundamental issue of how to know which lessons truly are, or are not, applicable to other situations.

This issue is non-trivial, particularly given evidence that even the most controversial technologies—nuclear facilities, genetically engineered organisms, incinerators—do not evoke uniform societal responses conceptually, when applied at particular locations, or over time. Nuclear power plants, verboten in the United States for nearly three decades, have been sited in Europe and Japan. Will the taboo on them in the US ever lift? Similarly, the use of agricultural biotechnology is pervasive in the US but, until recently, an anathema in Europe. What, then, constitutes a lesson?

We suggest that it is at least as important to understand what leads to variations in societal responses within categories of technologies as between them. Clearly, most technologies in and of themselves do not produce particular societal responses. Other factors or dynamics must come into play. Discerning the sets of conditions that lead to different kinds of societal responses should eventually enable researchers to generalize within and across technologies.

To what extent can we accurately anticipate societal responses and acceptability?

Stated differently, our first fundamental statement is "Here we go again. Another technology whose proponents believe is an obviously improved, though perhaps imperfect, way of addressing important societal needs fails to be embraced by society." Maybe the surprise is not anticipating controversy or rejection. But, our second statement is "Wait a moment, societal responses are not so simple." The same technology meets with different societal responses in different locations and at different points in time. These two statements lead to our question: "Given the apparent contradiction between the sameness of and differences in societal responses to new technologies, can we anticipate societal responses to new, rapidly emerging technologies?"

From an academic perspective, the ability to anticipate accurately marks a shift from fragmented case studies to scientific principles. In application, this ability allows decision-makers and planners to:

(a) improve their proposed activity by incorporating anticipated concerns into their design or plan, thereby also preventing or minimizing unnecessary conflict;
(b) identify situations in which their proposed activity is likely to meet with substantial controversy or conflict, regardless of its design or plan; and
(c) avoid inadvertently promoting conflict either by ignoring societal concerns, assuming incorrectly that what did or did not "work" elsewhere is applicable to their particular case, or operating on the belief that public education will lead to particular outcomes (an "if they only knew what we know, they'd agree" mentality).

How can, or should, society make better-informed decisions?

We suggest that the main reason for understanding and anticipating societal responses to new and emerging technologies is to contribute to better, and better-informed, decisions. At its simplest, "better" means that the decisions (a) halt or minimize adverse impacts of technologies and their applications and (b) accommodate the development and application of technologies that can produce substantial societal benefits. Not all new or emerging technologies "should" be adopted if they fail to produce net societal benefits, however those benefits are gauged. At the same time, it can be jarring to consider what might have been, had societal discomfort stifled the discovery of the double helix or the development of such transformative technologies as computers or cell phones.

Agricultural nanotechnologies— members of a class of technologies

We see agricultural nanotechnologies as members of a class of technologies that include other nanotechnologies, biotechnologies, cognitive technologies, information technologies, and such energy technologies as biofuels, hydrogen, fuel cells, and fusion. This class of technologies is marked by several attributes.

- First, they have—and are promoted as having—the ability to revolutionize science, technology, industry, or society. This hype fails to acknowledge the possibility that dis-benefits (costs, writ large) may be equally revolutionary.
- Second, the technologies are enabling in the sense that they can be used in disparate applications. Nanotechnologies, as an example, can be applied in agriculture, environmental remediation, medical diagnostics, pharmaceuticals, packaging, cosmetics, clothing, etc.
- Third, this class of technologies promises tremendous benefits at the same time as it poses substantial potential costs. These benefits and costs may be financial, environmental, or cultural, and can affect human health and well-being. Further, the distribution of benefits and costs is uneven; benefits typically are separated in space, time, and/or social class from costs.
- Fourth, these technologies are fraught with uncertainties and ignorance (the so-called unknown unknowns) about their efficacy and about their direct and indirect consequences.

- Fifth, this class is typified by science-in-progress. The underlying science and resulting technologies for any single member of this class (e.g. agricultural nanotechnologies) are immature and emerging at different rates. Thus, some products are deployed while others exist only as visions of what is possible.
- Sixth, science and technology research and development simultaneously are being conducted in academia, government, and private sector laboratories. Whether or not there are formal collaborations between or among these sectors, the historical distinctions among them are blurring. Universities and academics increasingly seek to generate money through licenses, patents, and spin-off companies in ways that once would have diminished their standing as independent, unbiased parties. Lines between government and industry can become fuzzy with increased pressure within government laboratories to commercialize. In short, the stereotypes of independence in academia, concern for public welfare in government, and quest for profit in the private sector are breaking down, making it more difficult to trust projections and analyses of societal implications from any of the three.
- Seventh, these technologies provoke value conflicts. Synthetic biology, genetic engineering (especially transgenics, where genetic material is transferred from one species to another), and nano-biotechnology clearly instigate value conflicts by raising questions about what is human vs. non-human or natural vs. non-natural. These technologies, like others in the class (including agricultural nanotechnologies), also prompt value conflicts over what constitutes appropriate boundaries delimiting the circumstances in which they should or should not be used. Take, as one of many possible examples, radio frequency identification (RFID) tags or quantum dots. Should their use in agriculture be banned, limited to circumstances such as tracking cattle to help prevent or quell the spread of diseases like mad cow or hoof and mouth, or unlimited, allowed in uses that include food products purchased by consumers?

Classifying technologies as "emerging" is more than a matter of convenience or curiosity. Rather, it is an important step in starting to answer systematically the fundamental questions we raised at the beginning of this chapter.

Patterns of societal response can be anticipated

We hypothesize that patterns of societal response can be anticipated. We conceptualize three broad categories of societal response to emerging technologies:

- clear-cut;
- seemingly chaotic; and
- random.

Clear-cut responses may be deterministic and readily apparent. They therefore do not compel our attention. Because random responses by definition are not predictable, we do not concentrate on them. Instead, we focus on the category of societal responses that seemingly are chaotic. We hypothesize that structure—a system of rules—underlies this apparent chaos. We seek to understand this structure to be able to anticipate future societal responses. Our choice of the word "anticipate" is deliberate. Our intent is to identify predictable patterns of societal response, not to predict specific behaviors in specific situations. Therefore, in seeking to distinguish the generalizable from the unique, we seek to understand both the empirical and the conceptual bases for comparisons within and among technologies.

At present, these questions remain unanswered and the ramifications of this situation are significant, even calling into question the premise for this book. It seems patently obvious that lessons learned from agricultural biotechnology should be applicable to agricultural nanotechnology. However, two assumptions are embedded in this comparison. One assumption is that the appropriate basis for comparison is by economic sector (agriculture). The second assumption is that biotechnologies are comparable to nanotechnologies. While these assumptions may be valid,[4] other alternatives exist. Examples of other possible bases for comparison include the following:

- Boundary-crossing
  - transgenics; inserting/embedding technologies into biological organisms; synthetic biology
- Open vs. closed
  - dispersed via field (open-air) application vs. integrated into manufactured product; consumer manufacturing vs. laboratory setting; on-the-organism vs. within-the-organism application

- Geographic setting
  - rural vs. urban; region/place within the US; US vs. non-US
- Scale
  - spatial scale—localized vs. regional, national, or global
  - population scale—individual organism vs. populations of different sizes and compositions
  - temporal scale—short- vs. long-term or intergenerational
- Familiarity
  - simple substitute ("new and improved" clothing, cosmetics, packaging, food item, or other product) vs. novel or unfamiliar (self-assembling product or organism, square tomato)
  - historical precedent vs. unknown

There clearly are many dimensions along which comparisons could be made. But what is the theoretical or conceptual basis for determining which comparisons should be made and, by extension, for anticipating future responses? In the following section, we offer one conceptual framework for consideration. We envision it as an initial step toward the development of a robust theory of societal responses to emerging technologies.

Suggesting a conceptual framework: PACT

To address these concerns we have developed a conceptual framework called PACT (public acceptability of controversial technologies) through our research on the determinants of social acceptability of bioremediation technologies[5] (Wolfe and Bjornstad, 2002, 2006; Wolfe et al., 2002, 2003). Bioremediation, in the context of the subsurface metal and radionuclide contamination on which we focused, involves the use of microbes to stabilize the movement of contaminant plumes.

PACT is decision-oriented. This decision-making is complex from the standpoints of the contaminants themselves, the physical environment in which they are found, and social interactions and decision processes. Our emphasis has been on communities—collective responses, interactions, and decision-making—rather than on individuals or broad policies. Our interests have been in decision-making that entails a strong element of public involvement and the potential for generating controversy. Technologies or solutions that fundamentally fall within the private sector and represent marginal advances (e.g. "simple" substitutes like higher efficiency refrigerators), or are aimed at individual consumer choices, generally have fallen outside the purview of our consideration in developing PACT.

In part, PACT formalizes a decision-oriented concept of acceptability as a response to new technologies. We defined acceptability to mean a willingness to consider a technology or problem-resolution alternative seriously. "Acceptability" is quite different from "acceptance," which implies a choice; our notion of acceptability need not imply a choice. Moreover, acceptability need not lead to actual technology deployment because there can be many steps between a formal decision and its implementation.[6] In PACT, acceptability is determined by agents (e.g. individuals or organizations) and their attributes; social, cultural, and institutional contexts; bargaining options; technology attributes; and the characteristics of the decision-making process. Note that the same agents can play different roles—with different standing, power, influence, and authority—in different contexts or in different decision-making processes. As just one example, a national non-governmental organization may have an enormous influence within a local community through such venues as public meetings, blogs, letters to the editor, or advisory groups, but have no standing in a local public referendum.

Our early conception of acceptability emphasized fluidity for three main reasons. First, we saw acceptability as a continuum, rather than as binary. The anchors of this continuum are (a) absolute positions, whether absolute acceptance (there is no question that x is acceptable) or absolute rejection (there is no condition under which x is acceptable); and (b) utter indifference. Both anchor positions represent non-acceptability because they shut off discussion. However, acceptability lies between the anchors, where responses are conditional ("yes...if" or "no...unless") and the parties involved are willing to consider alternatives. Conditions range from the minor and easily controlled to major, not easily controlled.

Second, acceptability reflects shifting positions on an array of topics. Proponents and opponents need not center on the technology itself. Rather, the individuals or organizations that constitute the "agents" involved in decision-making attach different degrees of importance to varied issues. While some agents see technological attributes as most important, other agents may see financial costs and benefits as the strongest determinants of decisions. Still other agents care most about which decision processes are used, and the degree to which those processes are participatory. We posit that, for an individual or a group, overall acceptability can either be dominated by a single issue or be the result of interconnections among issues. Interactions among collections of these individuals and groups, each with their own proclivities with respect to acceptability, create the community-level decision-making dynamic. We are especially interested in dynamics that propel agents from conditional toward absolute responses. Because there is an asymmetry in the power of negative versus positive responses to technologies, information or interactions that propel parties toward absolute rejection are particularly important to identify and understand. Among the possible forces that act to propel parties toward absolute rejection is the divergence in primary issues on which acceptability is based. For example, imagine an interaction between parties for whom technological attributes are key and parties for whom participatory decision-making is key. Providing information or arguments centered around technological details inadvertently may antagonize and propel participation-oriented agents toward absolute negative positions.

Third, acceptability is fluid in that it changes over time. The shifting of positions just discussed constitutes one form of this change over time, where acceptability changes in response to new knowledge or interactions with other parties over the near term. Acceptability also may change over longer periods of time. First, some decisions are made over the course of many years. Environmental cleanup decision-making is one such example. Agents' initial positions may change considerably over that lengthy period of time. Second, some categories of decisions seem to emerge episodically, with alternating periods of quiescence and resurgence. Conditions may change substantially such that, with each revival, acceptability issues and dynamics change. Current attention to new ("new generation") nuclear power plant construction takes place within a different technological, economic, environmental, and political climate from discussions in the decades preceding or following the Three Mile Island accident.

Our initial formulation of PACT formalized an acceptability "system" (see Figure 8.1).

The acceptability "system" focuses on decision-making. The central arrow designates the extremes of acceptability outcomes—either binary (absolutely for or against) or totally negotiable; the rules or conditions that affect agents' placement or movement along that continuum; and the informal and formal decision-oriented dialogue

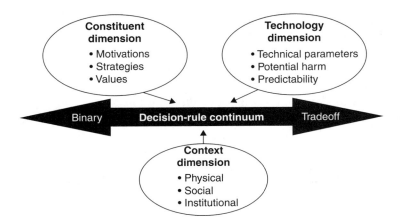

Figure 8.1   Schematic of the initial PACT conceptual framework

process. The "technology" dimension refers to the technology or option of concern, in the context of its alternatives. This dimension also includes a variety of technological attributes (risk, scale of application, predictability, etc.) that are brought to bear on the decision process. "Constituents" are the agents involved in decision-making, whether individuals or groups. This dimension encompasses the motivations, strategies, and values of the parties involved in decision-making.[7] And, the "context" dimension includes physical (e.g. hydrogeology, vegetation, land cover), social (e.g. size of community; economic dependence on the proposed technology or solution or on the organization proposing those options; community cohesion, etc.), and institutional (e.g. nature of proposing institution; local, state, and national regulators; trust/distrust of proposing institution, governmental organizations, etc.) elements as they have operated over time.

Later, we extended our conception of this PACT system to emphasize linkages across technologies or problems of interest and linkages over time. Linkages across technologies can be thought of as "spillover" effects; they encompass the ways in which choices about related technologies can affect choices about the technology of concern. This volume's emphasis on what agricultural nanotechnology can learn from agricultural biotechnology formalizes this linkage. Other potential spillover effects are embodied in questions such as what will become the next "frankenfoods" and whether responses to nanosensors could affect responses to smart treatment delivery systems.

Linkages over time refer to choices early in a technology's life cycle that can affect options later. How technologies and issues are framed early on can open or close downstream options. This process is illustrated by how boundary-challenging and boundary-crossing issues are defined. For instance, are nanoparticles to be regulated as "substantially equivalent," very small versions of bulk materials or as new materials with different properties? Whether this question is answered affirmatively, as seems to be the current trend, or negatively affects how nanomaterials are regulated and treated in downstream applications.

Technologies have life cycles, starting with basic and/or applied research, and progressing through demonstration, deployment, decommissioning, and disposal phases. Each life-cycle stage reflects different concerns and creates different downstream impacts. As examples, during basic science stages, health-effect concerns primarily center on laboratory workers and the communities surrounding laboratories. When technologies are deployed, different populations on local through global scales may be affected, sometimes voluntarily and sometimes involuntarily. Affected populations may or may not have influenced deployment decisions, and may or may not benefit from the technology development or deployment. And, disposal may affect local through global populations—including future generations—who may not have been parties to decision-making and who may not benefit directly from the technology.

Whether or not PACT proves sufficiently robust as a framework for understanding and anticipating societal responses to emerging technologies like agricultural nanotechnologies, we believe that there are several attributes such a conceptual framework should embody. First, it should provide a systematic, comprehensive, and dynamic approach to acceptability that provides a means to draw distinctions within and among technologies. Second, it should focus on the conditions that influence acceptability and the outcomes of acceptability decision processes, not simply on opinions or the diversity of those opinions (at particular points in time or over time). It should ask "under what conditions is x technology acceptable," rather than "to what extent is x technology acceptable." Third, the framework should incorporate a strong inter-temporal dimension. This temporal dimension should allow for responses to "shocks" to the system such injections of new information, introductions of competing technologies, and repercussions of key events. It also should be sensitive to the evolutionary or life-cycle stages of technology development, demonstration, deployment, decommissioning, and disposal.

Conclusion: a call for a convergent science of societal response

We have decades of experience with a range of rapidly emerging technologies. Still, there currently is no systematic, comprehensive, and dynamic framework to provide guidance for answering fundamental societal response questions, anticipating societal concerns, or analyzing comparisons across experiences or technologies. For those engaged in decision-making about emerging technologies like agricultural nanotechnologies, it is difficult to know what kinds of societal responses to expect. And, expectations—whatever they may be and whatever the bases for them may be—often collide with reality. Societal responses may seem random rather than chaotic, with underlying rules.

A framework that helps to distinguish the unique from the generalizable can help to move us out of this morass by moving us along the path leading to accurate anticipation of societal responses. We offer PACT as a nascent theory that describes the behavior to be studied; incorporates evolutionary aspects of emerging technologies; and leads to a descriptive, and eventually anticipatory, conceptual framework. Among PACT's attributes are that it was created by members of different disciplines and that it deliberately seeks to provide a holistic perspective that does not adhere rigidly to any single discipline. Many of the potentially revolutionary, emerging technologies represent a coalescing or converging of different disciplines. Likewise, we suggest that it is time to develop a convergent social science to study and anticipate societal responses to emerging technologies like agricultural nanotechnologies. No single discipline, alone, can assess and address these complex issues adequately. A convergent science of societal responses can move away from disaggregated case studies and uni-dimensional analyses, toward a coherent suite of underlying principles that govern behavior. Perhaps, then, the déjà-vu experience will result from an anticipated pattern of responses rather than from being surprised over and over again.

Endnotes

1. The submitted manuscript has been authorized by a contractor of the US Government under contract No. DE-AC05-00OR22725.

Accordingly, the US retains a non-exclusive, royalty-free license to publish or reproduce the published form of this contribution, or allow others to do so, for US Government purposes.
2. http://www.brainyquote.com/quotes/authors/y/yogi_berra.html
3. Note that most of these technologies are the subject of decision-making in which members of the public are involved to varying degrees. But, with the exception of energy-conserving technologies, this involvement does not entail individuals' purchasing decisions since the products are not consumer goods.
4. There is no proof that these assumptions are *not* valid.
5. This research was funded by the Natural and Accelerated Bioremediation Research Program, Bioremediation and Its Social Implications and Concerns Program Element, Biological and Environmental Research, Office of Science, US Department of Energy (Grant KP1301010).
6. Formalized decisions to accept a technology or solution (e.g. via a public referendum) are not always implemented for a variety of reasons, such as difficulties in securing financing or obtaining necessary permits.
7. Note that we did not include within the "constituents" dimension normative considerations about who *should* be involved in decision-making; disaffected and powerless parties who are not engaged in the decision process directly or indirectly thus are not included within this dimension.

References

Keller, K. H. (2006). Nanotechnology and society. *J Nanoparticle Res* **9**, 5–10.

Roco, M. C. (2005). The emergence and policy implications converging new technologies integrated from the nanoscale. *J Nanoparticle Res* **7**, 129–143.

Shatkin, J. A. and Berry, B. E. (2006). Approaching risk assessment of nanoscale materials. *NSTI-Nanotech* **1**, 253–256.

US Environmental Protection Agency (2007). Nanotechnology White Paper. EPA 100/B-07/001.

Wolfe, A. K. and Bjornstad, D. J. (2002). Why would anyone object? An exploration of social aspects of phytoremediation acceptability. *Crit Rev Plant Sci* **21**, 429–438.

Wolfe, A. K. and Bjornstad, D. J. (2006) Making imperfect decisions: Results from public workshops on bioremediation. *Applied Anthropol* **26**, 54–64.

Wolfe, A. K., Bjornstad, D. J., Russell, M. and Kerchner, N. D. (2002). A framework for analyzing dialogues over the acceptability of controversial technologies. *Sci Technol Hum Values* **27**, 134–159.

Wolfe, A. K., Bjornstad, D. J. and Kerchner, N. D. (2003). Making decisions about hazardous waste remediation when even considering a remediation technology is controversial. *Environ Sci Technol* **37**, 1485–1492.

Internet references

Hart Research Associates, Inc. (2006). Report Findings. Based on a national Survey of Adults. September 19, 2006. Accessed April 5, 2007. http://www.wilsoncenter.org/index.cfm?fuseaction=topics.home&topic_id=166192

Renn, O. and Roco, M. C. (2006). Nanotechnology: Risk Governance. International Risk Governance Council. Accessed April 5, 2007. http://www.irgc.org/irgc/projects/nanotechnology/_b/contentFiles/IRGC_white_paper_2_PDF_final_version.pdf

Savage, N. (2006). EPA: Nanomaterial risk evaluation extramural research activities. Accessed April 5, 2007. http://www.hesiglobal.org/Committees/ProjectCommittees/Nanomaterials/Nanomaterial+Presentations+091506.htm.

Areas of Ambiguity in Implementing an Emerging Technology

A Framework for Translating Biotechnology Experiences to Nanotechnology

9

David Sparling

With few exceptions, the process of developing and distributing new technologies to society is a commercial activity motivated by profit. The impact of new technologies cannot always be understood from a scientific perspective—there may be significant social and business implications. Biotechnology innovations have reshaped the agricultural, food, and health industries, resulting in the redesign of business models, business operations, and the structure of industries adopting them. The impacts of nanotechnology promise to be even more widespread and profound, affecting industries which are based on physical, as well as biological, sciences. However, the experiences with

What Can Nanotechnology Learn from Biotechnology?
ISBN: 978-012-373990-2

biotechnology offer an opportunity to anticipate the challenges and opportunities associated with nanotechnology and prepare both industry and society for the changes which could occur.

This chapter examines the commercial aspects of new technology development and introduction, and implications for the development of nanotechnology. A framework for understanding the new technology introduction process within a broader societal context is developed using the experiences of biotechnology. I consider how this framework could be applied to nanotechnology and the implications for both business managers and policymakers.

New technologies from discovery to market

The innovation process for converting scientific discoveries into technologies or treatments that benefit society is illustrated in Figure 9.1.

The process is a development and transition from a focus on science to a focus on products and markets. The process begins with a scientific discovery, often in a university or research institution through publicly funded research. A scientific discovery is assessed relatively early to determine whether it has sufficient economic potential to warrant the expenditure of time and resources needed to commercialize it. At this point, a link is hypothesized between the

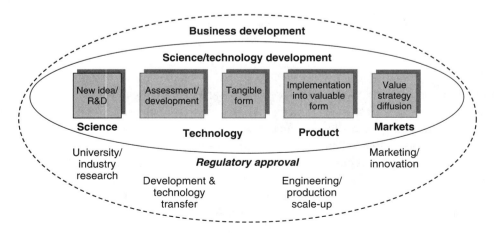

Figure 9.1 Innovation process

discovery and potential products destined for identified markets. If the markets are too small the project may be abandoned or relegated to the status of research to be continued for its scientific appeal, not for its commercial application. One of the critical elements in most new technologies is the ability to acquire intellectual property (IP). Without IP firms may have no ability to capture the benefits of their investment in a new technology, because it will be copied by others. In the United States, biotechnology research aimed at disease treatments were typically funded by the National Institutes of Health (NIH). Investment in agricultural biotechnology was different. Universities and research institutions were heavily involved in the development of the basic science. However, the real push for agricultural biotechnology came from private companies who made the connection between the potential of biotechnology to create crops with herbicide-resistant or insecticidal properties and the massive global market potential for such products. The promise of multibillion dollar markets encouraged firms like Monsanto to invest from the early 1980s to commercial launch in the early 1990s.

From science to technology

If the discovery holds promise then the process moves to the next stage, converting the discovery into a technological reality, one that is tangible and real, but not necessarily in final marketable form. While some of this development may be undertaken within research institutions, generally they lack the resources and the incentives to complete the transformation of a discovery into a technology prototype. Thus public sector technologies must be transferred to industry partners to complete the development. This will only be done if the technology meets several criteria:

1. Market size—The target markets for the technology must be large enough to warrant investing millions of dollars over a period of years.
2. Technological advantages—The technology must exhibit significant technological advantages in efficacy or efficiency over existing technologies in or those predicted to be in the market.
3. Technological feasibility—Developing the technology must be within the capabilities of the acquiring firm and the time and cost

to development must be in line with the expected market size and market advantages.

4. Ability to capture profits—The firm must have the ability to capture the value in its investment, either through IP protection or some other strategy.

5. Fit with firm business strategy—The technology and its markets must fit the strategy of the acquiring firm. Note that in some cases, the strategy of the acquiring firm is to continue technology development to a near market state and initiate the regulatory approval process and then to license the technology to a large firm to scale up production, complete regulatory approval and take the product through to the market.

It is at this second stage that firms take their first scan of the market opportunities and potential obstacles they might face. Biotechnology firms tend to sell their products, whether they are medicines or crops, to customers who are not the final consumers. In the case of agricultural biotechnology, developers of genetically modified crops viewed farmers as their customers and they correctly assessed the attractiveness of the products to those customers.

Products and markets

The third stage involves turning the technology into a product which has value to a market or markets. This stage is a complex combination of technology refinement, securing regulatory approvals and production scale-up in preparation for the commercial launch of the product. Once the products and production are ready, the final stage involves diffusion of the technology to initial target markets, supported by a marketing and distribution system. If the launch is successful the emphasis will move to supporting the product, continued product refinement and expansion into new markets. It is at this stage the firms attempt to capture the value of their investment. It is also the time when unforeseen problems with market acceptance can adversely affect a new technology.

The links between the different parts of the process are inextricable. The influence of the anticipated final products and markets reaches back to affect the earlier stages, defining the nature of the science and technology development and the path taken to market. Decisions made early in the process can have major impacts on the characteristics of the final product and its reception in the markets.

Radical technologies and innovation

When we consider the impact of new technologies we have to look beyond the technology to the innovations in the processes employing the technology and in the organizations that adopt or are affected by the technology. An innovation may be defined as a change to a product, process, or organization (or some combination thereof) that is new to the organization implementing it (Boer and During, 2001). An innovation must provide value to the organization (Tidd *et al.*, 2001), value that may be measured in economic, environmental, or social benefits. Innovation is different from invention. Invention is the creation of a new idea, while innovation is the application of that idea. A single invention can lead to many innovations as different operating units adopt and or modify the invention. For example, the creation of techniques to splice genes into plant DNA would be defined as inventions, while the development and deployment of new biotechnology crops are the resulting innovations. Innovations may be broadly divided into two classes depending on whether they apply primarily to the *technology* of an organization or to *organizational* structure and operation.

Technological innovation

Technological innovations alter a firm's products or processes. In Figure 9.2, we classify technological innovation on two dimensions: first by whether the innovation applies to a product or process and second by the degree to which the resulting product or process differs from current products or processes.[1]

We define products to include both physical and service offerings to customers. Process definitions are equally broad, including all activities and technologies involved in the production and delivery of a good or service. As innovations frequently include both product and process attributes, we categorize them on a continuum of a product/process mix. The mix varies depending on a firm's competitive strategy and the product's position in its product life cycle (Gopalakrishnan and Bierly, 2001).

Innovations are also classified by the degree of change they cause. Most innovations are incremental; they are logical and progressive improvements to existing product or processes. Radical innovations represent a major divergence from the existing situation and generally result from major scientific or technological breakthroughs, in this case genetic modification. Radical innovations can change the basis of competition in an industry and dramatically alter its structure.

Organizational innovation

Innovation can also occur in the structure and processes of organizations or institutions. We categorize organizational innovation, which includes changes to institutions, on two dimensions (Figure 9.2). The first deals with the timing of organizational innovation relative to product or process innovations. Innovations in institutions, structure or management processes can drive technological innovations, evolve simultaneously with technological innovations, or result from adapting to technological innovations.

The second organizational innovation dimension deals with the extent to which changes are internalized within the firm. While some innovations, like total quality management, may be primarily internal, others, such as supply chain management and traceability, cannot occur without extensive involvement by network partners.

Innovation and agricultural biotechnology

The relationship between technological and organizational innovation depends on the extent of change involved in each. The introduction of a radical technological innovation will inevitably lead to significant organizational change and there will be a shift between the two as the system moves toward a new equilibrium, albeit one that keeps shifting but not as dramatically. The key innovations in the introduction of agricultural biotechnology are portrayed in Figure 9.2 and positioned according to the characteristics of the innovation.

The innovations related to agricultural biotechnology occurred throughout the supply chain.

Innovations at the genetics level

The first truly successful genetically modified (GM), herbicide-resistant and Bt crops, were radical product innovations resulting from

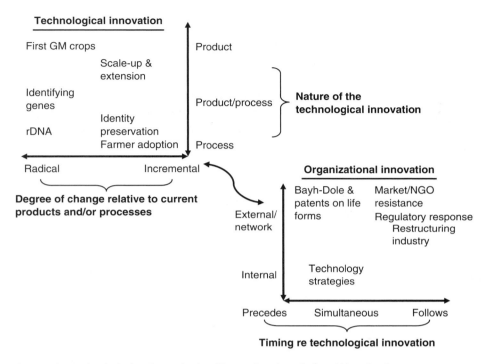

Figure 9.2 Technological and organizational innovations in agricultural biotechnology

years of research and massive investments by industry, academic and government research centers.[2] Genetic modification of plants required several innovations: identification of genes to be inserted in the plants, development of radical new processes to insert genes into plants, and incremental innovations to tissue culture methods to grow modified cells into plants (Evenson, 2002). The resulting biotechnology crops were effectively bundles of innovations. The development was facilitated by two institutional innovations: the Bayh-Dole Act of 1980 which allowed academic institutions to own and profit from their inventions and the issuance of patents on modified life-forms. Organizations could now protect their inventions and reap the benefits of their investment, providing greater incentive to invest in biotechnology research.

As biotechnology competence improved, firms made incremental improvements to extend desirable genes into other crops, with new varieties and future developments supported by research into plant genomes.

Implementing the innovation

Agricultural biotechnology provides an excellent example of how radical technologies can alter an industry. Biotechnology changed the relationships among agricultural chemical firms, seed suppliers, and their customers, linking seeds with herbicides or by reducing the need for pesticides. Producer purchase decisions focused on single genes rather than the performance of the entire genetic packages. The perceived benefits of the Roundup-resistant and Bt genes were such that they provided Monsanto with sufficient market power to demand that producers sign technology transfer agreements,[3] an organizational innovation allowing Monsanto to capture additional rents and to dictate production controls to maintain seed effectiveness. Roundup Ready crops provided an additional benefit to Monsanto in that they extended the life of Roundup beyond the expiry of the patent for its active ingredient.

Innovations at the seed company level

Seed companies traditionally developed their products through long-term breeding programs based on combining entire genetic sets. The advent of biotechnology meant that single genes were now highly valued property. These firms had to secure access to the new genes and integrate them into their varieties and hybrid seed lines. Seed companies had difficulty reorienting themselves to this new environment. At first they struggled to deal with biotechnology firms, but later, as the value of their germplasms was recognized, the industry was restructured as seed firms were acquired by biotechnology firms like Monsanto, Novartis, and DuPont who needed productive plant platforms for their genes.

Innovations at the production level

Inventions do not become innovations until they are commercialized. Farmers were quick to see the benefits of these new crops and the diffusion of GM crops was rapid. Farmers made incremental process innovations to adopt GM crops, changing production processes to accommodate the new technologies. These innovations resulted in tangible benefits in labor, yield, and/or quality, with few changes to post-harvest crop management.

Innovations and the grain and oilseed supply chains

When GM crops were first introduced, no significant changes were anticipated in grain distribution or use. The industry assumed that the final product was equivalent to unmodified crops. Resistance to GM crops in Europe changed that view. Environmental organizations and consumer advocacy groups made incremental changes, cooperating and repackaging their activism to resist GM crops on the basis of environmental and health concerns. Resistance to GM crops led to their rejection in many European retail chains and to reformulation of some products to eliminate ingredients potentially containing GM ingredients. Grain and oilseed supply chains to selected markets had to create new identity preservation systems for their non-GM supply chains and employ new testing technologies, from quick strip tests for field use to labs dedicated to testing GM products.

Governments in the European Union, Japan, and other regions implemented policy innovations to better reflect consumer wishes, placing moratoriums on introduction of new GM varieties and establishing labeling regulations for GM content. North American governments and industry groups responded with initiatives such as the Council for Biotechnology Information to promote GM crops and to minimize consumer resistance.

New technologies and industry structure

The restructuring of the agricultural seed, biotechnology and chemical industries as a consequence of agricultural biotechnology has been well documented (Hayenga, 1998; Charles, 2001). The need to secure plant and biotechnology IP resulted in the purchase of smaller agricultural biotechnology firms by companies such as Monsanto and DuPont. Later, agricultural and pharmaceutical biotechnology firms sought to merge to build on perceived synergies between their technologies, creating the life sciences strategy. Ultimately, resistance to GM crops and negative publicity around agricultural biotechnology forced the decomposition of life sciences companies into agricultural and pharmaceutical components to maintain shareholder value. Firms also realized that expertise in one area of biotechnology did not necessarily translate to other areas.

The introduction of agricultural biotechnology was an exciting but tumultuous period in the sector. Faced with a radical new technology, managers, consumers, and policymakers all struggled with how to respond to the technology. Looking back we can see a number of lessons which may be useful to policymakers and managers in developing new nanotechnology products.

Where did agricultural biotechnology go right?

New technologies are almost always developed because of the business opportunities they offer. Governments support them as a means of advancing science in general and to create economic and social benefits. Firms have a much less complicated mandate. They are motivated by profit, since profits allow them to survive and to achieve other goals. In the case of agricultural biotechnology, the firms correctly recognized the profit potential in the multi-billion dollar markets for herbicide- and insect-resistant crops. They invested in products that they knew their farmer customers would purchase and where they could capture much, but not all of the value created through their IP. In other words, they invested in technologies that would provide profits and increase shareholder value, technologies that warranted a prolonged investment in areas from R&D and regulatory approval to production and acquisitions of genetics companies.

Although there have been challenges, from a business perspective it is impossible to characterize the introduction of GM crops as a failure. The decision to focus on input traits was a good one in terms of both regulatory approval and producer acceptance. The advantages were easily discernible to producers and the fact that resulting crops were deemed "substantially equivalent" to existing crops greatly simplified regulatory approval. Biotechnology companies managed to introduce new GM crops into North America with little fuss at the consumer level but achieved remarkable market penetration at the producer level in a very short time. The use of technical user agreements was not overly popular with producers at first but was accepted once farmers recognized the value of the crops to their businesses.

Where did agricultural biotechnology go wrong?

Managers and investors in virtually all areas of biotechnology underestimated the time and money required to get their products to market. The rule of thumb of twice as much money and twice as long to market often mentioned for developing new technologies was not enough for most early biotechnologies. The first biotechnology company, Genentech, had its initial public offering in 1980 but did not market its first biotechnology product under its own label until the early 1990s. Similarly, Monsanto began research into genetic modification of crops in the early 1980s but did not launch its first products until 1992.

When they launched their products, agricultural biotechnology companies understood their customers and their needs. Their mistake was that they did not understand the final consumers and ignored them. Because GM and non-modified crops were "substantially equivalent" from a scientific perspective, managers and scientists assumed that their acceptance in the marketplace would be substantially equivalent as well. In North America, this was generally true, partly because consumers trusted regulators more than EU consumers did and partly because they remained largely unaware of the presence of GM crops in their foods. When consumers in the EU and Japan in particular balked at GM crops, managers in agricultural biotechnology firms assumed that they just did not understand the science and thought that if they could enlighten consumers the problem would be solved. They were excited by the potential offered by the science and assumed that they had not been clear enough in conveying that excitement. They also viewed the non-governmental organizations (NGOs) as troublemaking environmentalists who were attacking GM crops as a means of promoting their organizations. While this may have been true in a few cases, the view did little to help the companies manage the challenge to GM crops.

The excitement regarding the technology also made many companies ignore the importance of the complete genetic package; the genes must be inserted into high-yielding varieties. Similarly, plant genetics companies used to working with entire genetic combinations underestimated the value of single genes, leading many to ultimately be taken over by biotechnology companies, who had greater access to cash at the time. A technology is not usually a complete product. In the case of crops, the Roundup Ready or Bt genes had to be put into high-producing varieties. The inability to understand the importance of different varieties in tomatoes ultimately led to the failure of Calgene's Flavr Savr tomato and the company behind it.

Another error made by biotechnology companies in general was the so-called "life science" strategy, the view that if a company was good at one aspect of biotechnology it could buy or merge with companies in other areas to create super biotechnology companies that could do it all, from agriculture to medicine. With the technology boom overheating markets, biotechnology companies overpaid for acquisitions or merged to create life science companies. Synergies proved elusive and when agricultural biotechnology fell out of favor and stock values suffered, agricultural biotechnology companies were spun away from firms focused on medical applications.

What can nanotechnology learn from agricultural biotechnology about the business of launching a new technology?

The development path, public reaction and impacts on industries depend on the products introduced and the markets they serve. The actions of major industry players, governments, and constituent organizations can influence both industry development and public reaction to new technologies. Looking at the experience with agricultural biotechnology there are several very clear lessons for developers of nanotechnology:

1. New technologies destined for commercial markets have to make economic sense. The product/market mix has to be clearly defined and the potential payoff has to be in balance with the development risk.
2. Plan for far longer and more expensive development and regulatory approval times.
3. Look at the entire product package. How much of it depends on the technology? Where will the rest come from?
4. Understand the target markets. Understand the benefits that the technology brings to participants in those markets. What motivates the customers and consumers? Are the two different? The consumer is the most important part of the chain. How will the technology change what and how they buy?
5. Consider the impact of introducing the technology on the entire supply chain. How will it affect players at each level of the chain? How will it alter business models and business processes? Where might there be resistance and why? In agricultural biotechnology, the challenge came from the consumer end of the chain. With other technologies, it might come from another level.

6. As business models and processes change, the companies and institutions using them will adapt. If the technology is radical, expect and anticipate structural and institutional innovation. Being proactive can smooth transitions for companies.
7. Focus is essential in new technology development. Being a high performer in one area of the technology will not necessarily translate to all areas. Involvement in too many aspects of the technology hurts company focus and confuses both customers and investors.

Introducing new technologies is not simply about the technology and its advantages. It is the result of the complex interplay between technological and organizational innovations and markets. Disruptive technologies like biotechnology and nanotechnology change the basis of competition and, as a result, the structure of the industries they serve. Carefully considering all implications and impacts of different nanotechnology product/market combinations can help policymakers and managers better understand the likely result of the introduction and avoid some of the challenges that occurred in the introduction of agricultural biotechnology.

Endnotes

1. This discussion is an adaptation of the work by Tidd *et al.* (2001, p. 8), who categorized innovations into product, service, or process.
2. Charles (2001) provides an excellent summary of the development of agricultural biotechnology.
3. Technology transfer agreements specify terms to which a producer must agree in order to purchase seed.

References

Boer, H. and During, W. (2001). Innovation, what innovation? A comparison between product, process and organizational innovation. *Int J Technol Manage* **22**, 83–107.
Charles, D. (2001). *Lords of the Harvest.* Perseus Publishing.
Evenson, R. E. (2002). Agricultural biotechnology. In: Steil, B., Victor, D., and Nelson, R., eds. *Technological Innovation and Organizational Performance.* Princeton University Press, pp. 267–284.

Gopalakrishnan, R. and Bierly, P. (2001). Analyzing innovation adoption using a knowledge-based approach. *J Eng Technol Manage* **18**, 107–130.

Hayenga, M. L. (1998). Structural change in the biotech seed and chemical industrial complex. *AgBioForum* **1**, 43–55.

Tidd, J., Bessant, J., and Pavitt, K. (2001). *Managing Innovation*, 2nd edn. Wiley.

Engagement and Translation: Perspective of a Natural Scientist

10

Hans Geerlings and Kenneth David

Nanotechnology has become one of the most highly energized disciplines in science and technology preceded by biomedical research and defense. Increasingly, "nanotechnology" is heralded in the media as the "next big thing" and companies like Shell are approached by universities and venture capitalists with request to join (that is, to help fund) "nanotech consortia" or "nanotech start ups." Briefly speaking, nanotechnology involves the ability to "manipulate atoms and molecules" and it deals with devices with atomic or molecular scale precision. In general, devices with dimensions less than 100 nanometers are considered products of nanotechnology (1 nanometer $= 10^{-9}$m). The field is a vast grab bag of stuff that has to do with creating tiny things that sometimes happen to be useful. Nanotechnology borrows liberally from condensed matter physics, engineering, molecular biology, and large swaths of chemistry. Researchers who once called themselves materials scientists or organic chemists have transmuted into nanotechnologists. These nanotechnologists have been able to adeptly capture and hold public attention—in this case the votes of lawmakers

What Can Nanotechnology Learn from Biotechnology?
ISBN: 978-012-373990-2

in the US and EU who hold research purse strings. (Statement by Scientific Researchers of Shell Global Solutions International BV)

Let us start with two case studies (Case studies 10.1 and 10.2) describing working with tiny things that have turned out to be useful and that have not wrecked the environment—even after a protracted period of industrial application.

These two case studies are a useful contrast to the fictional picture of nano-processes as unbridled, irresistible, destructive forces. These cases also respond to the frequently voiced fear that the impacts of nanotechnology will only become apparent years later.

This chapter does not argue that all uses of nano-level particles are safe. We hold that the very small proportion of the US National Nanotechnology Initiative (NNI) budget set aside for all risk assessments is unacceptable (for details, see Chapter 1). We shall argue that

Case Study 10.1

"Slicing" platinum: making better use of precious metal resources

1. Platinum is an expensive material. It is highly useful as a catalyst that reduces energy of activation of chemical reactions.
2. Because particles of platinum adhere to one another, very little surface area is available for catalytic use.
3. You cannot simply "slice up" platinum in order to increase surface area. As soon as you do, it will begin to "sinter", that is, agglomerate into larger particles.
4. To prevent platinum from sintering, you first make supports out of porous solids (made, for example, from sand—silicon dioxide).
5. Next, you create nano-scale particles of platinum within this porous support. The support prevents the small particles from moving and thus prevents them from sintering. One method is to dissolve platinum salt in water and then introduce this solution to the support. Excess water is driven off by heating. Other methods exist for supporting the catalyst.
6. Supported platinum catalyst now has a huge surface area. One gram of platinum can result in roughly 300 m^2 of surface area. Twenty grams could cover a football (soccer) field!

Implication: this nano-scale process makes more effective use of precious resources, and reduces of energy of activation.

Case Study 10.2

From whale oil for lamps to zeolite catalysis: zeolite catalysis is a nano-level process employed for over 30 years without noticeable negative safety or environmental impacts

1. Whale oil, boiled from blubber, was the main reason for whaling. Used both in lamps and as candlewax, whale oil was the chief source of light in Europe and North America. Later, whale oil was used for oiling wools. "It was the first of any animal or mineral oil to achieve commercial viability" (Wikipedia, 2007).

2. Petroleum, using crude refining techniques, produced lamp oil; the rest of the crude oil (including gasoline) was discarded. This is an environmental pollution. It was also inefficient use of the resource. Early refineries in various parts of the world—Indonesia and western Pennsylvania, for example—discarded gasoline.

3. An improvement in refining techniques was "catalytic cracking" using natural materials. Minerals with a microporous structure are called zeolites. Picture, if you will, really tiny sieves made of clay. Petroleum was refined via various clays as catalysts at the nano level. This nano-scale process produced diverse products, including high-octane gas that helped Allied airplanes during World War II. This process also reduces the proportion of waste materials to utilized materials in the crude oil.

4. A further advance in refining technology was catalytic cracking of petroleum using human invented materials at the nano scale. Mobil Oil invented synthetic zeolite catalysis at the nano level about 30 years ago; this invention is now beyond patent and is freely used by other companies.

5. Key to this process is a supported catalyst with selective-shape porous material as the support. Reactant (petroleum) enters the pores and begins the catalyst-aided reaction called catalytic cracking ("cat-cracking"). Energy of activation is reduced. Different components are strained apart. In addition to reducing environmental pollution and making more efficient use of the resource, output products are more smartly targeted in the process.

6. Synthetic zeolites hold some key advantages over their natural analogs. The synthetics can be manufactured in a uniform, phase-pure state. It is also possible to manufacture desirable zeolite structures such as Zeolite A, which do not appear in nature.

(Continued)

> ### Case Study 10.2   (Continued)
>
> 7. Since the principal raw materials used to manufacture zeolites are silica and alumina, which are among the most abundant mineral components on earth, the potential supply of zeolites is virtually unlimited.
>
> Implications: this nano-scale process makes more effective use of abundant resources such as silica, avoids use of irreplaceable resources such as crude oil, reduces environmental waste, and, after 30 years of use, causes no recorded, hazardous effects.

the processes of assessment and of social acceptability regarding nano-level innovation need more attention. In this chapter, we focus on two particular issues in this area: issues of *translation* and *engagement*.

Focus

Issues regarding translation

Working on "tiny things that sometimes happen to be useful," faithfully reports the perspective of research scientists within Shell Global Solutions who work on nanoscience leading to nanotechnologies. This perspective contrasts with the usual critical statements about the mode of communication between scientists and the public. That is, in the lengthy public dialogue concerning biotechnology, critics represent scientific communication with the public as the "knowledge deficit" model: Scientists develop rational knowledge while the public is uninformed and irrational. Given this knowledge deficit, scientists teach the public what they need to know to gain agreement and compliance. In other words, scientific communication towards citizens is characterized as stratified, authoritarian, asymmetrical, unidirectional, and condescending (Toumey, 2006).

We can ask whether this construction better fits the mode of communication practiced by spokespersons for the scientists, corporate strategists, or policymakers rather than by research scientists themselves. One focus of this paper, then, is translation: translation of messages among four parties: scientists, resource allocators (company executives or research grantors), citizens, and governing agencies (standards-setting and regulatory agencies).

Issues regarding engagement

We shall rethink the continuing call for scientists to take a greater part in considering social, equity, privacy, ethical, legal safety, and environmental implications of their nanoscience. Scientists are asked to participate with citizens in more transparent dialogue on emerging technologies: Various methods of engagement such as consultation papers, small-scale citizen juries, and larger scale events ("GM Nation?" in the UK) have been tried. These events have been sharply criticized for alienating the public when they occur too late in the progression of innovation (progression from initial discovery of a scientific idea through research, design, development, and commercialization of an innovative product) (Rogers-Hayden and Pidgeon, 2006). Upstream public engagement is recommended: participation in decisions made well before the product is ready for market.

Calls for upstream, transparent, democratic participation are frequently heard in bodies such as the Society for Philosophy and Technology (where philosophers are dominant) and the Society for Social Studies of Science (where social scientists dominate). A question that needs to be asked is how much nanoscientists are in charge of the decision whether or not to consider such implications during their research? Who pays the piper? Who calls the tune? Whether in academia or in business, scientists' research occurs in the context of a system of decisions (proposing ideas, evaluating them, deciding on them, and then allocating resources to implement them). Resource allocators have the final decision concerning allocations of three resources (time, funds, and human resources). Scientists face go/no go decisions, whether from business strategists and research program administrators (in companies) or research grantors (in academia). Calling solely on scientists to allocate time for societal impact statements is therefore partially missing the point. What needs to be addressed is the relationship between the resource allocators and the scientists regarding the processes of engagement.

The purpose of this chapter, then, is to highlight the perspective of practicing research scientists: a voice that can go unheard in debates about emerging, controversial technologies (see Berne, 2005).

Two particular themes will be emphasized: the *timing of engagement* among the four parties and *translation of messages* among the four parties: specifically, among the scientific/technical community, resource allocators, the public, and governing agencies.

When is a likely time for scientists and the public to communicate? Our analysis suggests that a viable timing of engagement is achieved when three constraints are optimized:

- acquisition of sufficient knowledge to deliver reasonably reliable impact statements,
- credibility of the engagement, and
- competitive advantage of engagement.

In brief, while the first section of this chapter on the timing of engagement discusses the "when" of engagement, the following section on the translation of messages discusses "who," "how," and "what" of communication during engagement.

Note that this essay was collaboratively written by a natural scientist who specializes in research involving nano-scale processes and an organizational anthropologist who has modified social movement theory for research on acquisitions, joint ventures, and dispersed engineering projects, and the mobilizations by opponents and proponents of nanotechnologies.

Engagement

Definition of terms

Two key terms in the public debate on bio- and nanotechnologies regarding participation and communication are *upstream participation* and *public engagement*.

Consumer advocates and NGOs urge the advisability of upstream participation: very early participation of the public in the process of discovering, developing, and commercializing a new technology. From this perspective, early public engagement is a check against undesirable technological developments. Some key ideas used by the advocacy community are as follows:

Participation
- *Inadequacy of traditional risk assessment*: This leads to the likelihood of a hazardous outcome and the extent of damage that will occur. Such assessment should be augmented.
- *Upstream public engagement*: What is this technology for? Why should we employ this technology and not another? Who owns the technology? Who takes responsibility if something goes wrong? (Kearnes *et al.*, 2006)

- *Risk assessment as a social and cultural process*: This involves public discussion of values to be protected, analytic methods to be relied upon, and partners of scientific issues to be addressed (Kearnes *et al.*, 2006).

Methods of engagement
- *Mechanisms for involving the public*: Consultation papers, focus groups, citizen juries, and stakeholder juries.
- *Methods for evaluating mechanisms* for involving the public in policy formation (Rowe *et al.*, 2006).
- *Decision aiding techniques*: Stating objectives, stating options, and describing consequences.
- *Objectives of engagement*: Representativeness, transparency, and accountability and evaluation of methods of engagement that contribute to these objectives (Tansey, 2005).

In this community, scientific communication is characterized as stratified, one-way communication—"hierarchical, unidirectional, and condescending" according to Toumey (2006)—that is delivered after all meaningful allocations of resources have been made. For these reasons, events such as the government-sponsored public debate about genetically modified (GM) foods in the UK called "GM Nation?" have been reported to be ineffective.

We referred earlier to engagement among four parties, as opposed to the more common phrase, public engagement. This choice is intentional. We hold that engagement is indeed critical to processes of engagement of scientists, citizens, standards-setting agencies, and regulatory agencies. In order to formulate an action plan that is more likely to be implemented by the various stakeholders, we hold that engagement should consider both issues of risk to people, workers, and the environment and issues of business risk due to social risk of non-acceptance as well as regulatory risk of non-acceptance.

Timing of engagement

Further, timing of engagement is a critical issue. The aim of our analysis is to optimize the timing of engagement. A viable range for timing engagement is delimited by three constraints:

1. The reliability of risk assessment (sets limit on onset of engagement)

2. A series of go/no go decisions by resource allocators (whether business managers or academic research grantors)
3. The timing of upstream engagement with the public and pre-market approval with standards-setting agencies (sets limit on end of period for engagement).

On the other hand, working outside this viable range for timing of engagement entails risks, such as confrontation with consumer advocate groups and disruption of commercial operations and strategic disadvantage when standards follow directions taken by competitors.

Again, from the perspective of a practicing natural scientist who is concerned with oversight of risk and with social acceptance of ideas discovered, developed, and then commercialized, it is necessary to consider both issues that arise in the organizational environment (the company or academic organization to which you belong) and issues arising in the societal environment.

To analyze the viable range for timing of engagement among companies, citizens, and standards-setting agencies, we consider two progressions: *innovation sequence* and *innovation funnel*.

Innovation funnel

Not all innovative ideas reach the wider world. Rather, a series of decisions and resource allocations by companies or by research granting organizations successively trims a set of scientific discoveries to the final deployment of products or processes. Figure 10.1 illustrates the five stages through which innovation passes before it reaches the marketplace. The idea of the "innovation funnel" is quite old and is part of the working culture of Shell Global Solutions, for example. "That department has a well-stocked funnel", means "They have plenty of innovative ideas." As a source, Shell scientists refer to a book by Jan Verloop (2004) (and references cited therein).

1. *Prospecting*: scientific curiosity; discerning patterns and laws of nature. Finding opportunities.
2. *Evaluating opportunity*: technical plus business assessment chooses some opportunities and rejects others. Ideas that are accepted receive resources. Ideas are accepted when they meet standard business criteria:
   (a) Marketability and economics: likely to make money
   (b) Environment: no apparent harm to people, to social structures, or to the environment

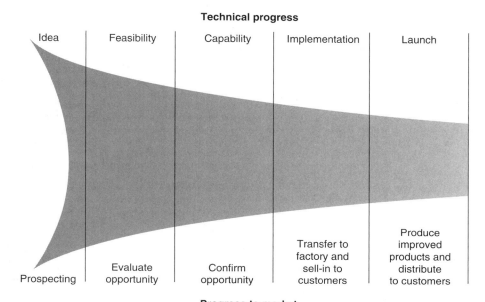

Technical progress

| Idea | Feasibility | Capability | Implementation | Launch |

Prospecting | Evaluate opportunity | Confirm opportunity | Transfer to factory and sell-in to customers | Produce improved products and distribute to customers

Progress to market

Figure 10.1   The Innovation funnel. From Russell Rankin *Food into Asia Program*, adapted from Ashok Ganguly, former director of Unilever

(c) Technological feasibility: feasible to produce.
Ideas are accepted also in terms of power competition among projects forwarded by different groups within a company.

3. *Confirm opportunity*. Market feasibility: strategic marketing plan is produced. Product prototype is developed and tested. Further technological development (designing methods and tools for producing the product and confirming its quality) requires further resource allocation.

4. *Fabrication planning, product and fabrication testing, and sell-in to customers (pre-launch marketing)*. Technical manuals produced.

5. *Product launch*.

This funnel of development also implies translations and hand-offs between functionally separate individuals or departments such as scientists, engineers, marketers, cost accountants, etc. All are needed to make a strategic business plan (formulate and implement a strategy). We return to the issue of translation later.

In summary, the innovation funnel (Figure 10.1) is a tool for locating temporal stages of innovation. It is also a guide to the series of resource allocation decisions that accompany an innovation.

Innovation sequence

The idea of an innovation sequence is variably defined by the public, by engineers, by business managers, and by scientists. A rather more detailed definition is helpful in order to define a viable range within the sequence—viable for timing of engagement.

To the general public, an innovation sequence can be described in a few terms: R&D (ideas are discovered and developed), production, and marketing. The R&D term actually puts together two terms that should be separated. Science is about curiosity. In this venture of learning, many things are allowed in order to learn how things work. Technology development is when you take advantage of what you have learned and seek to produce a practical, commercial application.

To engineers, the progression is more elaborate, involving discovery, design, development, demonstration, testing, and commercialization. Here, basic research is distinguished from product design, from development of technology needed to fabricate a product, from demonstration in the sense of building a working prototype, from testing the prototype, and, finally, from commercializing the product (Figure 10.2).

Engineering societies such as the American Society of Mechanical Engineers (ASME) have established environmental friendly guidelines termed the "cradle to grave" life cycle. This direction has also been expressed by Amy Wolfe, an anthropologist working on extra-technical concerns at the Oak Ridge National Laboratories: her definition of stages is as follows: basic research, applied research, demonstration, deployment, decommissioning, and disposal.

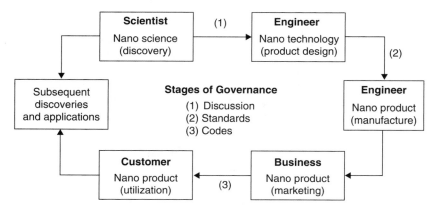

Figure 10.2 The product innovation chain (Source: Busch and Lloyd, Chapter 13, this volume)

To a business strategist, scientific/technical development, standards and regulatory approval, and business planning are all part of the process of innovation (Figure 10.3).

To scientists, the progression is more elaborate in specifying phases within basic and exploratory research.[1] The earliest phase is initial discovery of an idea. Initial discovery can occur via basic, curiosity-driven, fundamental research which most often is being done at universities, and via exploratory research as done in companies. The distinction is not absolute. The attitude in basic research is "how does it work?" The attitude in exploratory research is "how can I use it?" Scientists can shift between these orientations. In some companies, different departments segregate these activities. In both cases, such initial ideas are subject to review by resource allocators: academic research grantors or business managers. Research scientists' ideas are subject to a particular review in a company. They submit a one-page indication that the idea has promise for creating value in strategic directions prioritized by the wider organization. This statement results in a funding decision that reflects strategic and internal power issues and of course availability of funds. This statement, then, does not include enough information to spark a debate among scientists, let alone the wider public. By contrast, when research scientists submit ideas for review with a granting agency, the knowledge usually becomes public knowledge.

The second phase of basic or exploratory research may be termed global design in that the initial idea is subject to further specification.

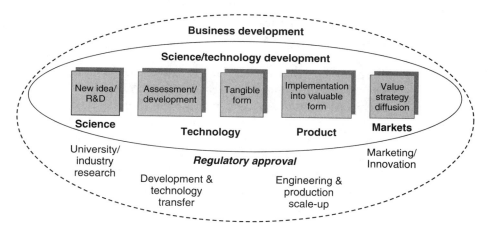

Figure 10.3   Innovation process (Source: Sparling, Chapter 9, this volume)

The idea is elaborated into a set of ideas that specify a set of processes that must be investigated.

The third phase, a detailed design, involves investigating the basic chemical or physical processes that are entailed by the idea. Research occurs with an experimental design.

Each of these three phases requires funding, but not much. Research colleagues at Shell estimate that out of the total cost of bringing a product to market only 10–30% is spent during the three phases of basic and exploratory research. From this perspective, we are dealing with nanoscience, not nanotechnology. At the same time, colleagues agree that if predictions about social, environmental, etc. impacts are made during these phases, the results are unacceptably unreliable. It is too early for discourse with the public because risk assessments based on findings from the stage of exploratory research are notoriously unreliable.

There is no scholarly consensus on the timing of effective public participation. Some scholars demand public discussion before any research is done. Other scholars such as Rogers-Hayden and Pidgeon (2006) state caveats: while early upstream engagement poses the possibility of influencing the trajectory of a future technology, there are definite challenges to upstream engagement when there is no specific product in sight and when citizen advocates are trying to ask new questions despite their lack of familiarity with the scientific findings. Our position is that early engagement during the period of basic exploratory research poses an unacceptable risk because scientific knowledge is not sufficiently reliable to present for reflection, scrutiny, and decision.

When, then, is the viable window for public and regulatory engagement? Entering the technical stage of product design and process design, we find the transition from research to application: the moment at which "the product/market combination" is defined in enough detail that public engagement is useful. Enough reliable knowledge is available for discourse. Public fora, citizens' juries, expert panels, etc. can be convened. As the cases presented in various chapters show, it can be costly for a company if they ignore the risk of public acceptance.

It is also a period when it is very useful for resource allocators to know whether or not to invest the really large amounts required for development and commercialization. In view of the resource allocation to be made by the company, it is still early days. Seventy to ninety per cent of investment looms ahead, the variation being dependent on the number of complex process steps that need to be demonstrated in the course of development. In addition, resource allocators should not wait too long for pre-market approval by standards-setting agencies. We coin the term "governance risk" to call attention to the fact that delay can have an adverse strategic consequence: a competitor can complete a pre-market approval before your company and achieve a competitive barrier via standards.

Some companies have taken a different line based on the informal policy, "Innovate. Then market your innovation." At a recent Nano4Food conference (Atlanta, November 13, 2006), a US Department of Agriculture official said "Get your product ready to market—then hire an expert marketer." There are consequences to this policy. As has been thoroughly documented in UK studies of public participation concerning biotechnology, late engagement leaves the public with the perception that all decisions have already been made. Late engagement—as in the "GM Nation?" debate in the UK—alienates the public.

Summing up, it is good for the public to participate at an appropriate time—when something tangible is available for discussion with the public. It is good for the company to learn at this point if what they are doing is actually too risky for commercialization. This reduces the likelihood of confrontation with consumer advocate groups and avoids having standards set by competitors! Let us restate our conclusion with a qualitative "formula" that companies could use as a project management tool for evaluating financial performance for new product development.

The following "formula" modifies a standard old project financial performance evaluation formula:

Projected profitability (that is, net present value of the project) = projected revenue/[cost of capital (%) + governance risk (%) + social acceptability risk (%)]

In financial analysis, the term risk is neutral, implying both positive and negative variations from the mean. In business policy analysis, environmental assessment includes assessment of both threats and opportunities for competitive strategy. Cost of capital, of course, is a precise amount. The other two terms in the denominator are intended to spur consideration and estimation of threats and opportunities that affect financial performance. If you fail to think about governance and social acceptability risks, you have no way of knowing whether your investment is going to pay off.

Do companies expect their scientists to display some alert recognition of social and environmental acceptability/hazard? In this region of policy, companies differ markedly in reward systems that actually drive personal decisions. Does the company reward people significantly (salary, stock options) for producing a major business success, a "killer application"? Such a reward system can prioritize the financial maxim (maximize stockholder value) and dull alertness to hazard and social acceptability. The choice of an internal reward system is entirely within company control. The results of such activity, however, may face external governance in the marketplace. Companies can and have decided that the cost of avoiding hazard (changing the design of gas tanks in Ford's Pinto) is less than the estimated cost of liability suits (Nader, 1965).

Conclusion re: timing of engagement
Our analysis of timing of engagement optimizes upstream engagement of scientists with public and governance processes on the one hand and reliability of risk-assessment statements and standards-setting processes on the other hand. In addition, timing of engagement is most practicable in the sense of fitting with the series of go/no go decisions that accompany technological innovation. A viable period of engagement is defined by the constraints of the innovation funnel and by the innovation sequence.

Phases of basic exploratory scientific research run concurrently with a series of business decisions granting relatively small budgets that forward or halt a particular innovation. These phases occur before sufficient knowledge is available for public reflection, scrutiny, and input. At the end of these phases, it is in the interests of all four parties (scientists and engineers, resource allocators, governing agencies, the public and its advocates) to engage in effective practices for participation. Sufficient knowledge should be available to permit reflection, scrutiny, input, and pre-market approval.

Resource allocators benefit by reducing social acceptance risk and governance risk. Companies can incorporate triple bottom line accounting as a normal procedure in the innovation trajectory. The public and its advocates benefit by reducing fear of hazard and alienation from participation events that they perceive to occur after all significant decisions have already been made. Governing agencies participate in concord with other parties and enhance credibility.

We can locate our position in terms of the views expressed by these other stakeholders of biotechnology and nanotechnologies, as shown in Table 10.1.

Table 10.1	Strategic advocacy choices regarding timing of engagement	
Citizen and environmental advocacy—upstream engagement	Advocacy of timed engagement of citizens, NGOs, government agencies, and companies	Profit advocacy— downstream engagement
Oversight of scientific and technological activities that maximizes upstream public engagement. Knowledge of social, environmental, ethical, legal impacts of a technology must be considered	Engagement scientific/technical community, public and business that optimizes: (1) upstream engagement of scientists with public and governance processes and (2) timing of engagement that valorizes reliability of impact statements	Research, design, development, and marketing that maximizes autonomy of the business enterprise and shareholder wealth. Only scientific findings are relevant in determining the risk of a technology
Precautionary principle	Triple bottom line accounting.[a] Consider financial outcomes, social, and environmental impacts	Innovate and then market the innovation—hire an expert marketer
Go? Only with transparent oversight	Balance discovery and innovation with oversight	GO for it! Maximize shareholder value

[a] Triple bottom line accounting proposes an alternative for business executives. This alternative is somewhat akin to the precautionary principle. In practical terms, triple bottom line accounting usually means expanding the traditional company reporting framework to take into account not just financial outcomes but also environmental and social performance (Elkington, 1998).

   In addition to timing of engagement there are further issues concerning the general question, "How to be more effective in engagement between scientists and the public?" We next turn to issues of translation.

Translation issues

Defining translation and boundary-spanning communication

Critics of scientific communication call for scientists to communicate in a more timely and effective manner. In summary, our position regarding effective boundary-spanning communication is that scientific knowledge must be translated in order to reach multiple audiences. The existence of multiple audiences implies that there are different communication boundaries to be spanned. Thus, various translations are necessary to achieve effective communication.

First, translation involves the recognition that a communication event is a relationship:

Sender $\rightarrow$ Message $\rightarrow$ Receiver

Second, more specifically, various components exist in a communication event. Meaning is constructed not only in terms of the content of message transmitted from sender to receiver, but also in the timing of the message, the medium through which the message is transmitted, and the mode of communication (style) (Figure 10.4).

Third, translation requires that the sender learn what is important to the receiver. If the content of the message sent both to an environmentalist and to an unemployed labor workforce is the same, the message may fall on deaf ears. The sender of the message must analyze the different priorities of the two audiences.

Fourth, based on this knowledge of audience priorities, the sender makes choices among options in each component and assembles distinct communicable messages for each audience.

During research in Sri Lanka, K. David and his research assistant visited the village of Chunnakam where a civil rights action was taking place. Members of an untouchable caste were seeking entry to a local temple. Entry was opposed with violence. Upon returning the home research village, a stream of visitors appeared to find out what was happening in Chunnakam. The research assistant gave a series of accounts. K. David later asked him why the accounts differed so significantly. The research assistant replied, "Each of the visitors has somewhat different interests and priorities. I just told each of them a version that reflected what they find important." (Field notes, Kenneth David, Ethnographic research in Jaffna, Sri Lanka, 1968–1970)

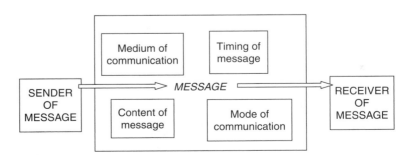

Figure 10.4 Components of the communication event

Note that this process of translation is tantamount to constructing a social relationship between the sender and the receiver.

The act of translation calls attention to an activity that accompanies scientific and technological innovation: the need to be an effective communicator. More specifically, to be a communicator capable of spanning diverse boundaries that may hinder the implementation of an innovation. Scientists and engineers are not typically trained in communication requisite to this task.

Effective boundary-spanning communication can become part of scientific education and thus contribute to the innovation process. Note that this includes a framework that brings together the variety of relationships that require boundary-spanning communication.

Communication tasks for scientists

In the longstanding debate about biotechnology, there is a history of regarding the scientific/technical community as communicating with the public in terms that are stratified and unidirectional and condescending: the "knowledge-deficit" mode (see Chapters 3, 5, and 11). Without questioning the occurrence of this knowledge-deficit mode of communication, one can still ask whether many scientists actually communicate directly with the public. Are program managers, business managers, and public relations people actually the spokespersons for scientists? When messages are delivered by scientists themselves, is the message carefully prepared by public relations managers?

While some scientists are indeed excellent communicators (see Box 10.1), communication deficiency is also recognizable.

A different sort of knowledge-deficit exists on the side of scientists and engineers—knowledge and capability as communicators. There is indeed a large literature on educating scientists and engineers on communications. But how effective is educational practice? Over the last decade, Kenneth David, an organizational anthropologist, has worked with John R. Lloyd, a professor of mechanical engineering, on a research and education program with engineers who are working in geographically dispersed engineering design teams—such as occur in outsourcing projects. Engineering programs produce technically competent engineers, but they are not educated in responding to communications, cultural, and power issues that directly impact on collaborative activity and project performance. When educated in transcultural

---

Box 10.1: Scientists as mass communicators

Not every scientist fits this characterization of stratified and condescending communication. Nobel prize-winning physicist Richard Feynman, addressed "the problem of manipulating and controlling things on a small scale" in a 1959 lecture "Plenty of room at the bottom," at the California Institute of Technology (Feynman, 1959). This lecture was of course an early step in the move towards the study of nanotechnology. Another version of the lecture was delivered to high school physics students. Feynman also wrote the entertaining and enlightening book, *Surely You're Joking, Mr. Feynman! (Adventures of a Curious Character)* (Feynman *et al.*, 1997).

---

capabilities (capabilities to recognize and respond to communications, cultural, and power issues) performance in dispersed teaming improves markedly (David and Lloyd, 2003).

The term "transcultural" here refers to crossing whatever barriers are necessary, including barriers in communications among departments within an organization or across organizational lines, communications within a country or across national lines, communication with the public directly or via public relations officers, communicating directly or indirectly with standards-setting and regulatory agencies. Figure 10.5 depicts the set of relationships relevant in the scientific/technological innovation process; the set of relationships surrounding the scientist (identified as "product and technology innovator") for which different communications are needed:

I. Relationships between the science/technical community and the public; communications are modified, augmented, and transformed both by mass media and by NGOs

II. Relationships among companies in a supply chain. Supply chain constraints impact on technological development

III. Relationships between standard-setting and regulatory organizations on the one hand and companies in the supply chain on the other hand

IV. Relationships among scientists, engineers, business managers, etc. in the organizational environment.

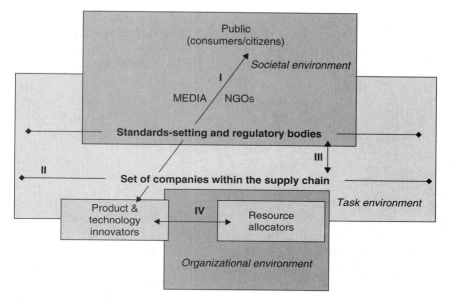

Figure 10.5   Engagement, supply-chain, governance, and resource-allocation

Note that this figure takes account of the relationships presented in the different perspectives of the business manager, the engineer, and the scientist presented earlier (see Figures 10.1–10.3).

In all of these cases, messages are translated because they are being communicated to quite different audiences. In various professional bodies such as the Philosophy of Technology and Society and the Society for Social Studies of Society, communications between the scientific community and the public receive much attention. We take the position that the scientist communicating with the public (**I** in Figure 10.5) is only one of the communication tasks of the practicing scientist.

Translations needed for boundary-spanning communications
In this section, we review varieties of communication with different audiences identified in Figure 10.5. Some audiences are internal to the organizational environment; others are present in the wider task and societal environments. Our core argument is that engagement requires not manipulation of meaning but practices for effective

translations: crafting of communication events that gets meaning across to the chosen audience. Recalling Figure 10.4 on components of communication, it is necessary to consider a particular target audience (the receiver of the message) and craft each component—the message, medium, mode, and timing of the communication event—so that the message is more likely to get across to the audience. We now consider each of these components of communication.

Diversity of messages

Messages differ greatly in the degree of detail related by the sender to the receiver of the message. Citizen advocates urge transparency in communications by scientists. In practice, scientific findings become somewhat lost in translation. Communications are restricted not only for proprietary advantage, but for feasible understanding. If you have something that is relatively complicated and you try to be informative, you must avoid overloading the receiver of the message. When a natural scientist (actually, any scientist) communicates, there is a filtering in the sense of a selection and reduction of details. A variety of filtering or translating processes occur:

- When you are presenting your findings to natural scientists or engineers who are research colleagues.
- When you are justifying a research endeavor to a resource allocator: to a business manager or executive or a research-granting agency.
- When your message is handled by your public relations department and they filter the details.
- When you address a class of graduate students, high school students, etc.

Even among peer scientists or engineers, transcultural communications have an impact. In Case study 10.3, the degree of reduction of detail mirrors the social relationships involved.

Diversity of media

Relevant media of communication include written texts, oral discussions, non-face-to-face electronic media, etc. The perceived meaning

Case Study 10.3

Communication among colleagues from different cultures

A major North American pharmaceutical company [Pharmco] set up a strategic alliance with a Japanese company. Chemical engineers from Pharmco began to send summaries of data on clinical tests to their counterparts. The Japanese engineers replied that they would prefer the full set of data, not summaries. Somewhat offended, the Pharmco engineers asked why their engineering colleagues did not trust them. The Japanese colleagues replied as follows: "We have not met you. We have not sat together in a conference. We have not gone to dinner together. We have not taken drinks together. We have not played golf together. How do you expect us to trust you?"

The Pharmco International V.P. arranged a series of visits in the United States and in Japan. With the social relationship and professional standing established after about 14 months, the Japanese engineers were quite willing to accept data summaries.

of messages very clearly changes according to media employed. Conventions for use of media are shaped by the national culture, professional culture, and organizational culture of the users (Terpstra and David, 1991). Organizational cultures as well as national cultures have different conventions for the use of media. Case study 10.4 illustrates that media conventions are shaped by different organizational cultures. In the incident, Americans from different companies miscommunicate due to different media conventions.

Transcultural communication, therefore, can occur even within the same city.

A cross-border relationship involves people from different national cultures. They can have rather different ideas on what it means to use a particular medium, as illustrated in Case study 10.5.

Different modes of communication

Messages can be delivered in different modes of thought. According to Bruner (1985), there are two modes: *paradigmatic* and *narrative*.

Paradigmatic thought is essentially the rational scientific approach, characterized by abstraction and generality. It is context-free, timeless

Case Study 10.4

Multi-media miscommunication between companies

Two companies were preparing for the annual renewal of a major contract. The companies are located in the same city in the United States. A pair of managers tried to arrange a business lunch. The appointment failed. They met the next day. "Why didn't you show up for lunch yesterday?" said the manager from one company. "You didn't invite me," replied the manager from the second company. "I certainly did. I emailed you three times to confirm the appointment." "Forget it. At my company, you have to telephone if you want to make a meal appointment."

A major contract was coming up for renewal and no more mistakes could be tolerated. The two managers reviewed conventions for communications and telecommunications within the contract negotiating team.

Case Study 10.5

Launching a multi-cultural project

In a Whirlpool Emerging Technology project involving US, Dutch, and Chinese branches of an engineering team, the branches wanted to use different media during project launch because they had contrasting cultural ideas about how one should launch a project.

The project objectives were not well defined by the client who wanted to "empower" the dispersed team to think innovatively.

How did the various teams respond? The US and the Dutch teams were task-oriented during the project launch: they worked hard to define the project and they communicated predominantly by email to express their ideas. The Chinese, by contrast, responded to "under-definition of objectives" by making a strong effort to establish harmonious social relationships with their counterparts until the time that project became better defined. They preferred to use telephone or Netmeeting for this purpose.

and universal. It seeks to establish truth through verification procedures and empirical proof. In this mode, action is understood in terms
of causes and correlations.

By contrast, the narrative mode of thought is based on story-telling.
It does not seek to establish truth but truth-likeness or verisimilitude.
It is not timeless but temporal, and it sees action as a result of human
intentions. (Gaskill, Chapter 12, p. 196)

Put more briefly, paradigmatic thought prioritizes value-free
observation, abstraction, and proof that convinces the audience.
Narrative thought prioritizes the evocative whole image that per-
suades the audience.

To these modes, we add the *WIFM mode*—as in "What's In It For
Me." WIFM mode may be compelling to the audience even when
they are neither convinced nor persuaded. Lee Dahringer, former
Dean of the College of Business, West Virginia University, related
an incident of WIFM in West Virginia concerning nanotechnology.
He reported that communicating with out-of-work people from
West Virginia is simple: "Just show me the job. Former coal miners
are used to danger."

In summary, as any effective communicator knows, learning to
understand the priorities of the receiver of the message is the first
step; then the communicator can work out the message to be delivered, the appropriate media for transmitting the message, the timing
of the communication event, and the mode of the message.

The following sections illustrate these points.

Case of multiple translations

Basic research on hydrogen storage was carried out by Shell
Global Solutions, a research unit within Shell of the Netherlands.
H. Geerlings worked on this area. Applications of this research, in
the form of public buses running on hydrogen, have been successfully launched in various cities in the Netherlands and abroad. The
messages to various audiences that were necessary to research,
to develop, and to implement this innovation—with regulatory
acceptances and with acceptance by citizens—are presented in Case
study 10.6.

In summary, the hydrogen case illustrates clear contrasts in the
entire communication event: different audiences receive diverse

Case Study 10.6

Research on hydrogen storage by Shell Global Solutions

Audience: the public

Message

The basic message for the public informs citizens with an account that stresses the practical value and the lack of risk of the innovation. The following topics provide a general introduction to the area:

- *Why do you want hydrogen-powered vehicles?* Hydrogen is an energy carrier with enormous potential for future energy provision, the more so since it is readily produced from a wide range of fossil and renewable sources. You will reduce local pollution. The by-product of a hydrogen-powered vehicle is just plain water. You also reduce global pollution if hydrogen manufacture is done the right way. Hydrogen is an energy carrier, not an energy source; you need to produce hydrogen with an expenditure of energy. You avoid pollution if you do not produce carbon dioxide at the same time.
- *Why do we want to store hydrogen?* A major issue with hydrogen, however, is efficient and cheap storage, and considerable research in this area is going on worldwide. In particular, the introduction of environmentally beneficial hydrogen-powered cars has been hampered by the lack of a safe, compact, light and economic hydrogen storage system. If you want to use hydrogen-powered vehicles commercially, you must store hydrogen in quantity necessary to drive the vehicle for 500 km.
- *How do you store hydrogen?* Storage is best done as a new compound—as part of metal hydrides. (Rechargeable batteries are either nickel metal hydrides or lithium ion.) There is a volumetric advantage: can store much more hydrogen in metal hydride than in compressed gas.

Media

Research scientists do not generally communicate directly with the public. *Shell Venster* (*The Window on Shell*) is an internal company publication that is made available to the public. The public relations department has a role in communications. Television or newspaper reporters set up contact via the PR department, for example, telephone interview from Vara broadcaster; PR will join the conversation.

Mode

- Paradigmatic: Low
- Narrative: Mid
- WIFM: High

Audience: a class of high school students

Message and medium

Lectures are sometimes delivered to groups of high school students educated with the basics of several natural sciences. Students are given a more systematic presentation than the general public. Lectures to be filmed for such presentations may be written by scientists but delivered by an actor who reads the script.

In these lectures a general overview of the hydrogen area is given. In such lectures, the story-telling mode (about applications to the real world) dominates rather than the paradigmatic mode (intense technical discussion). Alternatively, a group of students interviews a scientist, after which they have to give a presentation to their peers.

Mode

- Paradigmatic: Mid
- Narrative: Mid
- WIFM: Mid

Audience: research consortia

Companies participate in research consortia with universities, governmental bodies, and other companies.

Message

The objective of the ACTS program in the Netherlands is to develop competitive advantage for Dutch companies in the context of a knowledge society. All partners in the consortium pay a fee. Findings distinguish between effective and ineffective technologies, commercially attractive and unattractive technologies. Generic technology is generated by all the partners. Findings are communicated to all the partners.

Medium

Companies are at the board level. They approve of proposals submitted by universities.

(Continued)

Case Study 10.6 (Continued)

Mode

- Paradigmatic: High
- Narrative: Mid
- WIFM: High

Audience: peer scientists and engineers in a related field

The peer group of research colleagues is mainly but not exclusively composed of scientists; some engineers are part of this peer group.

Message and medium

To this group a very technical and rather detailed account of the state of the art is delivered through specialized colloquia, extensive written reports, or articles. These communications need not be brief. Two or three hour colloquia are delivered. The specific subject of discussion is not fixed and can include, for example, determination of atomic structure, methods of preparation, or aspects of system engineering, depending on whether the research colleagues are pursuing basic research or more applied research.

The image of dispassionate scientific discourse is not always the case. Discussion can be contentious if there is a power struggle involved. Berube (2006) documents at length the acrimonious and downright rude exchanges between idea opponents in the field of nanotechnology: exchanges between Kim Eric Drexler and George Whitesides, and Richard Smalley. Earlier, the scientific community reacted with ferocity to the writings of Immanuel Velikovsky (especially *Worlds in Collision*, 1950), who speculated that Venus passed close by the Earth and caused the Great Flood.

Technical discussion can be contentious if there is a power struggle involved: Shortly after the merger of ABN and AMRO, two commercial banks in the Netherlands, the two infomatica departments were told to integrate operations. As the two information systems were incompatible, the two departments realized that only one system would survive. A civilized war was waged in which each tried to gain supremacy. On one occasion, they debated for two hours over the meaning of the word for stock [*aandeel*], a technical concept that is usually rather familiar to commercial banks. The point was to establish which department's definition would prevail. Power can incite people to create miscommunication.

Mode

- Paradigmatic: High
- Narrative: Low
- WIFM: Low

Audience: research manager

Research managers in general have a solid technical background.

Message and medium

Research managers receive an abstract of a scientific report that would indicate all necessary technical details. Communication could be in the form of short accounts (one-pagers) but also in the form of more detailed verbal or written reports. Discussion does not imply simplification.

Mode

- Paradigmatic: Low
- Narrative: Mid
- WIFM: High

Audience: business manager

Business managers may have some scientific/technical background, but this cannot be presupposed. Because, communication between research scientists and business managers crosses the boundary between units within the company, there are conventions for communication.

Message and medium

In the early phase of idea discovery (see above, p. 193), communication is in the form of occasional brief meetings and a short, crisp, written message that emphasizes the business aspects of technology. The scientist gives a "crisp" qualitative explanation of the technical achievement and possible consequences. "Crisp" means six lines that answer certain questions. Why are we doing this project? What is the contribution? Technical breakthrough? New contacts with other organizations or with critical people? Are we pursuing a direction that is one of the company's designated strategic areas? If so, you do not need much more explanation.

In later stages of fundamental and applied research, scientific and technological progress is reviewed in "dedicated progress"

(Continued)

Case Study 10.6   (Continued)

meetings. In addition, scientists produce a deliverable management summary. This addresses various audiences with various objectives:

1. Findings aim to be judged adequate for continued allocation of resources for a new project. The summary presents intermediate results to a business manager who is the sponsor of an existing program. Message should be crisp (pertinent, concise, and timely) responses to request for information. The summary delivers a financial "bottom line" message. The message is not a standard financial net present value message. Rather, the message is an option/values message: that is, showing that the project has future potential value; buying the option to enter particular market/product directions. You have a number of options; if one works well you pay to pursue all the rest.
2. The summary also presents results, news of positive or negative results that might be informative to peers who are developing adjacent technologies.
3. The summary presents results to an upper level of management to inform them so that they know what is going on elsewhere in the company. A research department can issue a confidential journal twice a year for this purpose.

Companies differ drastically as to backstage communication patterns. A large North American consumer products company's "memos" ensure an absolute top-down communication pattern. Royal Dutch Shell permits upward communication in moderation to the very top of the company. Such internal consultancy (i.e. news of something that deserves management attention) "is what you get paid for."

Mode

- Paradigmatic: Low
- Narrative: Low
- WIFM: High

messages delivered via diverse media that emphasize different mode of communication. The occurrence of diverse translations of what scientists know does not automatically imply the delivery of condescending, paternalistic misrepresentation. Certainly, scientists

are not educated in communication arts, and public relations officers do not typically control an in-depth knowledge of science.
Engaging the public with a combination of these capabilities and
delivering skillful, careful translations is an area of endeavor that
needs development.

Discussion

Critics characterize scientific communication with the public as
asymmetrical, condescending, communication that is delivered too
late for the public to offer meaningful input. Citizen advocates call
for upstream communication that represents an innovation more
transparently than is currently being done.

This essay agrees that more effective communication is necessary
and proposes suggestions on how to accomplish that objective.
Communicating means establishing a relationship between senders
and receivers of messages. Adjusting the content of the message (not
assuming the audience is ignorant; assuming the audience is not
informed) is only one step. Different media (technical reports, business memos, lecture and discussion, visual representations, etc.) and
different modes of discourse (analyzed in this essay in three dimensions—paradigmatic, narrative, and WIFM) should also be chosen to
establish the relationship. The previous section of this essay explored
the timing of communication that is viable and practicable.

Translation, whether in natural or in social science, should recognize the limits to communication. Paul Bohannon, an anthropologist, lecturing on the Tiv people of Nigeria became frustrated with
his inability to fully communicate the nuances of his story to a
group of anthropologists in England. He said, ironically, "Well, if I
could tell it to you in Tiv, you would really understand!"[2]

It is often said in linguistics that all translations are false. Every
word in a language is a unique combination of sense (meaning) and
sense (aural sensation). A translator must choose between faithful-
ness to the original in terms of meaning and retaining the beauty of
the original language in terms of sensation. Our term translation,
then, recognizes that it is hard to communicate the entire original
message in terms the receiver can fully understand. Our account of
addressing multiple audiences with different messages and modes
intends to work towards limiting miscommunication. Crafting of

communication events that gets meaning across to the chosen audience is perhaps best done with a process of mutual engagement, that is, the scientist who is willing to take the trouble to communicate effectively and the audience who is willing to move in the direction of the scientist. There is always a translation process. There can also be processes of back translation (translating the translation back to the original) to confirm that messages are accurately received.

In conclusion, the two sections of this essay have aimed to clarify two closely related aspects of communications that occur during the innovation process. The first section on the timing of engagement discussed the "when" of engagement, the second section on the translation of messages discussed "who," "how," and "what" of communication during engagement. In addition, we focus attention on a set of relationships towards which communications must be directed, as illustrated in Figure 10.5. Attention to these points aims to clarify gaps in the existing discussion of emerging technologies.

Endnotes

1. Additional details on communication during phases are reported in Case study 10.6 on hydrogen storage.
2. For another sample of pitfalls in translation, see Laura Bohannon's attempt to relate *Hamlet* to the Tiv people (Bohannon, 1971).

References

Berne, R. W. (2005). *Nanotalk: Conversations with Scientists and Engineers about Ethics, Meanings, and Belief in the Development of Nanotechnology*. Lawrence Erlbaum Associates.

Berube, D. M. (2006). *Nano-Hype: The Truth Behind the Nanotechnology Buzz*. Prometheus Books.

Bohannon, L. (1971). Shakespeare in the bush. In: Spradley, J. P. and McCurdy, D. W., eds. *Conformity and Conflict: Readings in Cultural Anthropology*. Little Brown and Company.

Bruner, J. (1985). Narrative and paradigmatic modes of thought. In: Eisner, E., ed. *Learning and Teaching the Ways of Knowing*. University of Chicago Press.

David, K. and Lloyd, J. R. (2003). Tools for organizational learning and organizational teaching: Learning and communicating about collaboration in dispersed engineering design projects. In: Klein, G. and Nemiro, J., eds. *The Collaborative Work Systems Fieldbook: Strategies, Tools, and Techniques.* Center for the Study of Work Teams, University of North Texas, Jossey-Bass.

Elkington, J. (1998) *Cannibals with Forks: the Triple Bottom Line of 21st Century Business.* New Society.

Feynman, R. P., Leighton, R., Hutchings, E., and Hibbs, A. R. (1997). *Surely You're Joking, Mr. Feynman! (Adventures of a Curious Character).* W. W. Norton.

Kearnes, M., Grove-White, R., Macnaghten, P., Wilsdon, J., and Wynne, B. (2006). From bio to nano: learning the lessons, interrogating the comparison. *Science as Culture* **15**, 291–307.

Nader, R. (1965). *Unsafe at Any Speed: The Designed-In Dangers of the American Automobile.* Grossman.

Rogers-Hayden, T. and Pidgeon, N. (2006). Deliberating emerging nanotechnologies in the UK and beyond. Session on Risk Perceptions and Social Responses to Emerging Technologies, Society for Social Studies of Society annual meeting, Vancouver, November 2, 2006.

Rowe, G., Horlick-Jones, T., Walls, J., and Pidgeon, N. (2006). Evaluating public engagement: The case of the 2003 "GM Nation?" public debate. *Proceedings: Values in Decisions on Risk*, Stockholm, Sweden, pp. 423–431.

Tansey, J. (2005). *Ships that Pass in the Night: Cultural Theory and Risk Management or Cloning the Pangolin.* Maurice Young Centre for Applied Ethics, University of British Columbia.

Terpstra, V. and David, K. (1991). *The Cultural Environment of International Business*, 3rd edn. South Western Press.

Toumey, C. (2006). National discourses on democratizing nanotechnology, *Quaderini* **61**, 81–101.

Velikovsky, I. (1950). *Worlds in Collision.* Macmillan.

Verloop, J. (2004). *Insight in Innovation.* Elsevier.

Internet references

Feynman, R. P. (1959). Plenty of room at the bottom. http://www.its.caltech.edu/~feynman/plenty.html

Wikipedia (2007a). Whale oil. http://en.wikipedia. org/wiki/ Whale_oil

Biotechnology, Nanotechnology, Media, and Public Opinion

11

Susanna Priest

Introduction

Can we expect the same media and opinion dynamics to characterize the introduction of nanotechnology as characterized the introduction of biotechnology? Comparisons between the two are common. Current attention to the level of public receptivity for nanotechnology, the emphasis on early attention to possible environmental and health effects, and the search to find opportunities for "upstream" engagement—that is, to find ways to give voice to public desires and concerns at an earlier point in the development process—all result from experience with biotechnology, particularly the genetically modified (GM) food debate (which was largely unanticipated within both industry and science policy circles).

While some advocates of public engagement in the formation of technology policy are sincerely seeking ways to improve deliberative democracy, others are likely more concerned with heading off—or at least identifying—"problems" with public acceptance. A fine line exists between public engagement activities and market research in many cases; the former are represented as being motivated by a desire to give ordinary people a say in how technology is developed, while the latter is motivated by a more instrumental desire to test the waters of public opinion with respect to what people will approve, accept, and purchase in a free-market economy. These are very much two halves of the same coin. At any rate, both of these motivations for "doing things differently" for nanotechnology have their roots in earlier biotechnology controversies. Certainly there are important things we can learn from biotechnology about nanotechnology and public opinion.

Yet nanotechnology and biotechnology are quite possibly more different than alike in terms of the technologies themselves, a circumstance obscured by the use of these two very similar-sounding umbrella terms to encompass, in each case, a remarkably diverse set of technologies. These technologies are united, in the nano case, only by their association with very small size, generally involving knowledge and capacities newly accessible through the recent development of novel instrumentation, and in the bio case, only by their association with human modification of biological structures and processes, often (but not always) using newly developed recombinant DNA techniques. Biotechnology variously refers to genetic modification of organisms (whether plants, animals, people, or microorganisms), stem cell research, cloning, xenotransplantation, and—in older and now less common usage—engineering artificial substitutes for human body parts (such as limbs or organs). Nanotechnology currently includes everything from the development of materials made of very small particles—with applications ranging from better sporting goods to targeted cancer drug delivery to the use of nano-scale processes in electronic microchip manufacture.

A crucial difference often ignored in the comparison is that biotechnology tends to raise ethical issues in the minds of ordinary people that are not raised (at least not as commonly or as insistently) for nano. Developments in nanotechnology are more rarely described as "playing God." Non-living material does not seem to invoke concerns related to culturally sensitive concepts such as food (GM), babies (stem cell research), or individuality (cloning). So we should not expect these two classes of technology (nano and bio) to behave the same, in terms of public perceptions and preferences. We should continue to expect, and look for, differences as well as similarities.

In this chapter the goal is to explore the relationship between media messages and public opinion for these technologies. In order to understand more about this relationship, three related issues need to be thought through: the nature of the categories (bio, nano) themselves; what contemporary media research is telling us about the dynamics of media effects; and how to think about "publics" for science in a post-deficit model era.

Problematizing the categories

These categories themselves (nano, bio) are less reflective of the nature of the diverse range of products they include and more reflective of the nature of advancements in the technologies that enable the underlying science in each case. Yet from a public perception perspective, these categories are not particularly "natural" and are therefore difficult to grasp. The use of the categories themselves may be more confusing than not; while ordinary people do seem to understand that nano is "small" and that bio involves "genes," it is possible that this superficial understanding might invite overgeneralization from one application to another. If gene therapy in humans is dangerous, then perhaps all biotech is suspicious. If nanoparticles in cleaning liquids are dangerous, then perhaps all nanotech is suspicious. This probably can and does happen. But contrary to what is sometimes assumed, ordinary people do not necessarily lump all bio and all nano together—they do make meaningful distinctions when presented with a variety of applications, as reflected in both early focus group research on nanotechnology (Priest and Fussell, 2006) and extensive opinion research about biotechnology extending over many years. Social scientists interested in understanding how people respond to these broad categories of "biotechnology" and "nanotechnology" should remember that non-scientists are not likely to have neatly defined cognitive categories that follow either of these labels.

In fact, one of the ways nano and bio are alike is in the remarkable range of applications subsumed under each label, apparently as much for historical, social, and political reasons that are still not entirely clear as for scientific ones. From a communication perspective, the term "nanotechnology" seems to have emerged primarily as a rhetorical or political strategy designed to leverage public and private financial support for what is often referred to as the "next wave" of scientific and technological development. Nations have vied with one another to become international leaders in particular subfields of nanotechnology (Nordmann, 2006), and research not previously conceptualized as "nanoscience" became so designated in order to be seen as a part of this "next wave." The term "biotechnology," on the other hand, may have arisen partly as a kinder, gentler (that is, more acceptable) way to refer to "genetic engineering." This is an important context for understanding that both nanotechnology and biotechnology are artificial social categories invented for social purposes rather than, strictly speaking, scientific "disciplines" or even "fields." Drawing the parallel between the emergences of these two sets of technologies, then, in terms of public reaction, is both tempting and somewhat misleading.

Public opinion scholars stress that illusions can result when people are asked their opinions about something they do not really understand or by the use of terminology in survey questions that can have different meanings and connotations to individual respondents (see, e.g., Bishop, 2004). These meanings and connotations can also change over time. It may even be the case that some measurable international differences in opinion regarding nanotechnology and biotechnology could be rooted in subtly different associations.

For example, it is quite possible that Europeans may associate "biotechnology" more closely with "GM food," and therefore give it a lower approval ranking than North Americans. Canadians may associate "nanotechnology" more with potential health applications than do people in the US, focusing their attention on a different set of risks and benefits than pertain to (say) sporting goods.

These are plausible, although completely hypothetical, examples. However, such differences, if real, would likely be associated with differences in emphasis within media messages.

Differences in the nature of public (media) discourse among different nations and regions can create a slightly different set of associations that in turn influence opinion statistics. This makes it difficult to sort out which intergroup differences are actually the result of cultural differences in response to the same technologies and which are the result of small but significant differences in associations with these broad umbrella terms.

Media and public opinion

Media researchers are generally united in the conclusion that public opinion does not follow directly from media messages. Today the so-called "magic bullet" theory of media effects that are strong, immediate, direct, and uniform has been almost universally discarded by media scholars, although it is still regularly reinvented by commentators seeking to blame a host of social ills (such as prejudice, violence, and so on) on media representations. These are important issues. People do learn things from the media, which can shape their behavior and their images of themselves and others in a variety of ways, both negative and positive. But audiences bring to their understanding of media messages their own personalities, experiences, values, priorities, beliefs, and interpretations. Furthermore, media messages also reflect, as well as shape, the experiences, values, priorities, beliefs, and interpretations of the social groups that create them.

Specifically with respect to technology, Mazur (1981) claimed some years ago to have demonstrated a (presumably causal) relationship between negative media coverage of a technology and negative public opinion. While his correlations undeniably show an association, it is just as plausible on the basis of correlational data like these that the media—especially if doing a good job—reflect and perhaps even foreshadow public controversies as that they somehow induce them. (For further discussion of the limits of this model and data challenging it, see Gutteling, 2005.) The relationship between media content and public opinion is an extremely complex "chicken-and-egg" relationship that is not well captured by correlational studies that do not take the nature of news-gathering processes or audience cognition into account.

However, and while available evidence does not provide much support for the idea that negative or positive public opinion directly follows from negative or positive media opinion, this is not to say that the media have no effects. The most consistently demonstrable short-term effect of media coverage is "agenda-setting," the idea that the media call attention to certain issues and not others, and that in the process readers and audiences are influenced in their thinking about what are the most important issues of the day. Mazur may be partially correct in that by calling attention to particular technologies, the media may be inviting readers and audiences to consider them controversial; news media are usually in the business of covering bad news more often than good, and this is what we routinely expect of them.[1]

"Framing" (sometimes called issue definition or second-level agenda-setting) is another well-established media phenomenon that has measurable effects on public thinking. Framing refers to the selection of some aspects of an issue, topic, event, or problem for emphasis. This is a necessary part of newswork; the media cannot publish or broadcast everything, but must by necessity publish and broadcast only highly selected portions of "reality," and in describing a particular event or issue, journalists necessarily make judgments about which elements people most need or want to know and then must fit these elements into a coherent account or story. However, framing is also influenced by news sources (including public information officers, public relations practitioners, scientific experts, and members of advocacy groups) who want, in turn, to influence public opinion.

Given the breadth of applications involved and the problematic character of both the category "biotechnology" and the category "nanotechnology," framing is a potentially important influence in both cases. Toward the more recent years of the agricultural biotechnology debate, as evidence of divided opinion on both sides of the Atlantic became more and more difficult to ignore, the promoters of this technology attempted to reframe the debate as one about liberating a set of tools that could end developing-world hunger and malnourishment, eradicate diseases associated with nutritional deficiencies, and put subsistence farmers around the globe on a sound economic footing even if their available land was of marginal utility. While many scientists working in this area are no doubt completely sincere in their hopes that their work will bring these changes about, this shift in the frame of the debate was primarily about influencing public opinion. In this respect it has been, at most, only partially successful. What "frames" will dominate nanotechnology coverage over the next few years remains to be seen.

Finally, longer term, media also have what are known as "cultivation effects" on our general perception of the nature of the world. Whether we believe the world to be dangerous and risky or safe and welcoming, whether we believe certain groups (scientists, capitalists, people of a particular nationality or ethnicity or gender) to be trustworthy or unreliable, whether we believe certain forms of social behavior to be common and acceptable or deviant—in short, our perceptions of social reality—are believed to be built up over long periods of time in part through exposure to the media, alongside influences from other social institutions. In the same way, reactions to new technologies are influenced by reactions to older technologies; if previous technologies consistently carried unacknowledged risks that we heard about on the news, then most likely the newest technologies will also. Ordinary people certainly use their knowledge of older technologies as an important resource for asking questions about newer technologies. But it is also important to restate here that ordinary people also make distinctions between one technology and the next; those who oppose GM foods are not necessarily those who oppose nuclear power, for example. Having said that, naturally people's experience with biotechnology—whether seen as positive or negative, how trustworthy the experts are believed to be, whether the technology carries more risks or more benefits, whether ethical issues have been addressed—will certainly influence their expectations for nanotechnology in a general way.

Given that people make distinctions among applications within these broad categories, however, the expectations they bring to bear are not necessarily as general as the problematic "nano, bio" categories. In thinking about the risks of nanotechnology-based drug delivery systems, for example, focus group participants in the US regularly refer specifically to their observations about the reliability of the FDA with respect to regulating other drugs (Priest and Fussell, 2006). While the analogies people use to grasp issues associated with new technologies are not always correct in every respect, everyday life experiences—often including information gleaned from the news media—are a crucial cognitive resource people rely on to make sense of the unfamiliar. This will be true of each specific application as it rolls out; while generalized expectations for technology based on previous experience will continue to persist in the background, making decisions about individual applications and products will often rely on more specific associations.

Another aspect of the relationship between media content and the GM food and broader biotechnology debates that does need to be stressed for understanding emerging nanotechnology is that media opinion has very often—too often—been mistaken by the policy community for public opinion. Thus the illusion persisted for years that people in the US were almost exclusively pro-GM food because much of the early media coverage in the US was industry- and researcher-driven and therefore positive (Priest, 2000). The nature of the US news system is that it is largely source-driven, especially in technically complex areas. Differences of opinion that existed in the US were not visible to policymakers; variations in survey data suggested a public that was not certain how it felt, not one that was becoming (as it turned out) increasingly polarized.

Conversely, some segments of the European press follow a different, more aggressive, less "objective" tradition. While dissent in the US remained largely invisible in the news, with some of the exceptions remaining confined to local issues without reaching national prominence (Priest and Ten Eyck, 2004), differences of opinion about GM foods in Europe came to the surface much more quickly, and may have been exaggerated in press accounts. On the one hand, for complex reasons of culture and geopolitics, it is true that European reactions to GM food technology were somewhat more negative, especially initially. On the other hand, impressions about the relative absence of dissent in the US and about its relative prevalence in Europe were partially illusions derived from different press traditions and styles of coverage.

At present, media coverage of nanotechnology is largely positive (Stephens, 2005) and both North American and European opinion leans toward the positive as well (Gaskell et al., 2005a; Priest, 2006). This does not mean these will stay this way, nor that one controls the other. It remains to be seen what impact emerging health and environmental issues that might be associated with nanotechnology and its products will have on this climate of opinion. But nanotechnology policy and industry organizations appear to be rightfully concerned that missteps at this stage of nanotechnology's "roll-out" could have long-lasting consequences; as a result, so far we have seen a healthy transparency in discussions of these risks.

Social constructions of "the public"

Imagining a public ignorant of the science involved allowed agricultural biotechnology's developers—notably major US seed companies, in particular the industry leader Monsanto Corporation[2]—to discount public concerns except in efforts to "educate away" these concerns with heavy doses of appropriate scientific information. This assumption is sometimes called the "deficit model" of science communication because it implies that negative reception needs to be combated by filling up people with appropriate scientific facts, thus eliminating the presumed deficit and assuring popular support. Empirical evidence suggests only a weak relationship between knowledge about science and technology and attitudes toward it, however (Priest, 2001; Priest *et al.*, 2003; Sturgis and Allum, 2004). In contrast, it is now widely recognized that perceptions of risk are more dependent on levels of trust in industry and in the regulatory watchdogs of government than on levels of knowledge. Transparently and effectively dealing with risk issues as they arise, rather than attributing questions about risk to ignorance of science, is likely to be a far better strategy for maintaining public confidence in the long run.

The rhetoric around nanotechnology is not entirely free of this "deficit model" assumption. Despite the increased emphasis on upstream public engagement and enlightened public outreach, the specter of the ignorant public, while certainly not so prominent as in the early agbiotech years, has not yet disappeared. Its relevant twenty-first century form is a public presumed living in fear of nanobots:

> Now in the millennial year 2000 the principal fear is that it may be possible to create a new life form, a self-replicating nanoscale robot ... able to be programmed not only to make another copy of itself, but virtually anything else. (Smalley, 2001)

This new "straw man" conceptualization of "what the public might think" about nanobot images, a conceptualization associated with the now-infamous Drexler–Smalley debates over how best to think about nanotechnology,[3] recalls early concerns with "what the public might think" when reading tabloid descriptions of "Frankenfoods" in the European press. Both strains of thinking assume powerful media effects in the absence of evidence for them, despite over 50 years of searching by media scholars. In fact, one recent study of nanotechnology opinion concluded that exposure to Michael Crichton's novel *Prey* about escaping nanobot hoards actually left readers more positive about nano rather than more negative (Cobb and Macoubrie, 2004). And on the basis of available focus group data, ordinary people generally fail to associate nanotechnology with nanobots at all, and on the relatively rare occasions when they do, tiny robots able to work on our behalf are more likely to be seen as a benefit than as a risk (Priest and Fussell, 2006).

In short, popular fears about nanobots seem to be much ado about nothing, but constructing a public that is believed to think this way is not without consequences. Focusing media discussion and public education efforts on combating such irrational fears may miss opportunities to understand and address what other areas actually concern people, which in the case of nanotechnology involve issues like job stability, information privacy, regulatory oversight, and access to benefits—rather than escaping nanobots.

One of the other wrong turns that can be taken in thinking about public opinion is the unexamined assumption that "the public" is a homogeneous mass, the very "mass" that is incorporated in the term "mass media." In fact there are many publics for any given issue, some consisting of stakeholders and some not, and each public has unique values and concerns. One failure of the deficit model is that it conceptualizes a monolithic public that is distributed along a single variable—presence or absence of knowledge. Adding the dimension of level of interest, Miller (1986) has popularized a further subdivision into "attentive" and "interested" publics for science, plus everyone else. These publics can be expected to pay more attention to news about science, and in Miller's view are the most important audiences for that news. However, subdivisions that are finer grained yet may be even more useful in an era in which "upstream engagement" is becoming a broadly accepted goal.

Another way of thinking about the "publics" for science and technology involves considering a number of additional dimensions (Priest, 2006). On the basis of exploratory statistical analysis, North Americans can be divided into five groups: "true believers" who assume science and technology are inherently benign and to be supported; "utilitarians" who are generally supportive but want to weigh risks against benefits for each technology; "moral authoritarians" who rely on ethical and political leadership for making such decisions; and finally "democratic pragmatists" and "ethical populists" who believe everyone should be empowered to make up their own minds, primarily on scientific or on moral grounds, respectively. Gaskell *et al.* (2005b) have extended a similar analysis to European publics. While the particular groupings represented here are reflective of the survey data on which the analysis is based and should not be "reified" as absolute distinctions, this research illustrates a way of thinking in terms of a richer variety of "publics" rather than a single "public" within which only one or two variables are of interest.

Discussion

Arguably, policymaking for science and technology is moving from a model based on elite pluralism (the assumption that only educated and informed elites need to be involved) to an approach more inspired by the idea of deliberative democracy (in which ordinary people are given more of a say). Moving toward more public engagement and understanding the broad variety of publics that exist for science and technology are steps in this direction. Market forces—the potential economic cost to industry and government of public rejection of technologies in which enormous amounts of money have been invested—likely contribute to this trend.

Varied publics have a range of hopes and concerns with respect to new technologies—not only in the areas of environmental and health effects (positive and negative) but also with respect to more purely social impacts such as employment shifts, information privacy, access to benefits, and distributional issues (that is, who gains and who loses, a dimension complicated by trends associated with economic globalization).

These nuances are lost in discussions focused solely on knowledge deficits. Ordinary people can have legitimate concerns about the social issues associated with genetic engineering and cloning without fully understanding the science of DNA. Similarly, ordinary people can ask reasonable questions about the social impact of nanotechnology without fully understanding the details of nanoscience.

Policy made as a result of broad consultation should be more stable and less likely to be followed by backlash from a chorus of voices that previously felt silenced. Media discourse therefore needs to capture the broadest possible range of voices on these issues. However, media are in turn dependent on sources: pro-technology voices from industry and government and (though less dominant) anti-technology activist groups tend to get the most attention because these are the groups seeking out that attention. Journalists may need to rethink their social responsibility to reach a broader range of publics with information from a broader range of sources on a broader range of issues in science and technology policy. This undoubtedly means engineers and scientists must rethink their role in society and become more engaged as well.

To date, nanotechnology is unfolding in a more transparent way that reflects a great deal of knowledge gained from the recent history of biotechnology's encounters with the public. Despite exceptions (such as the UK's Prince Charles' widely publicized alleged concerns over "grey goo"), public response to date seems generally measured. It may be that nanotechnology is simply less controversial for the cultural reasons suggested above, or it may be that to some extent improvements in media coverage and the current investment in better communication and outreach are actually helping to promote more reasoned discussion rather than polarization.

However, the problem with the deliberative model—especially for such broad categories of poorly understood technologies as we are dealing with in the "bio" and "nano" cases—remains the same as always. Everyone cannot take the time to be educated on every issue. Arranging for broad and informed public discussion of every public issue is likely impractical, some argue on economic grounds alone, and this is especially true where the issues involve complicated new science. Ultimately, elected representatives must still make many (or even most) decisions on behalf of others. No adequately developed mechanisms as yet exist for gathering up the results of deliberative experiments or focus group research and presenting them to policymakers. However, at least in the US, a patchwork of privately funded organizations is beginning to take on this role, and university-based social scientists with substantial Federal funding are beginning to augment their efforts. Albeit by baby steps, we seem to be making progress in figuring out strategies for keeping science and society in step.

Endnotes

1. While news media are run as businesses in most modern nations, the emphasis on "bad news" is not just a matter of gaining audience share or boosting circulation through sensationalism. Although this kind of motivation certainly exists, in fact we do rely on the media to tell us when things go wrong. Journalism that consisted primarily of "good news" would be failing in its responsibility to alert us to problems.

2. At the February 2006 meeting of the American Association for the Advancement of Science (which took place in Monsanto's corporate home town of St. Louis, Missouri, and was cosponsored by Monsanto as well), an employee of the company publicly acknowledged that

the corporation had "made some mistakes" in its early public relations efforts. This kind of statement was visibly absent in prior years' discussions.
3. See Baum (2003) for key portions of this debate.

References

Baum, R. (2003). Nanotechnology: Drexler and Smalley make the case for and against "molecular assemblers." *Chem Eng News* **81**, 37–42.

Bishop, G. F. (2004). *The Illusion of Public Opinion: Fact and Artifact in American Public Opinion Polls*. Rowman & Littlefield.

Cobb, M. and Macoubrie, J. (2004). Public perceptions about nanotechnology: risks, benefits and trust. *J Nanoparticle Res* **6**, 395–405.

Gaskell, G., Einsiedel, E., Hallman, W., Priest, S., Jackson, J., and Olsthoorn, J. (2005a). Social values and the governance of science. *Science* **310**, 1908–1909.

Gaskell, G., Eyck, T. T., Jackson, J., and Veltri, G. (2005b). Imagining nanotechnology: cultural support for technological innovation in Europe and the United States. *Public Understanding Sci* **14**, 81–90.

Gutteling, J. M. (2005). Mazur's hypothesis on technology controversy and media. *Int J Public Opin Res* **17**, 23–41.

Mazur, A. (1981). Media coverage and public opinion on scientific controversies. *J Commun* **31**, 106–115.

Miller, J. (1986). Reaching the attentive and interested publics for science. In: Friedman, S., Dunwoody, S., and Rogers, C., eds. *Scientists and Journalists: Reporting Science as News*. American Association for the Advancement of Science, pp. 55–69.

Nordmann, A. (2006). Invisible foundations: Herbert Gleiter and the contribution of materials science. Unpublished paper delivered at Nanoculture Seminar, University of South Carolina, April.

Priest, S. (2000). *A Grain of Truth: The Media, the Public, and Biotechnology*. Rowman & Littlefield.

Priest, S. (2001). Misplaced faith: communication variables as predictors of encouragement for biotechnology development. *Sci Commun* **23**, 97–110.

Priest, S. (2006). The public opinion climate for gene technologies in Canada and the United States: competing voices, contrasting frames. *Public Understanding Sci* **15**, 55–71.

Priest, S. and Eyck, T. Ten (2004). Peril and promise: news media framing of the biotechnology debate in Europe and the U.S. In: Stehr, N., ed. *Biotechnology: Between Commerce and Civil Society.* Transaction Publishers.

Priest, S. and Fussell, H. (2006). Nanotechnology: constructing the public and public constructions. Unpublished paper, University of South Carolina.

Priest, S., Bonfadelli, H., and Rusanen, M. (2003). The "trust gap" hypothesis: predicting support for biotechnology across national cultures as a function of trust in actors. *Risk Anal* **23**, 751–766.

Smalley, R. (2001). Nanotechnology, education, and the fear of nanobots. In: Roco, M. C. and Bainbridge, W. S., eds. *Societal Implications of Nanoscience and Nanotechnology. Final Report of the Workshop Held at the National Science Foundation, Sept. 28–29 2000.* National Science Foundation.

Stephens, F. L. (2005). News narratives about nano S&T in major U.S. and non-U.S. newspapers. *Sci Commun* **27**, 175–199.

Sturgis, P. and Allum, N. (2004). Science in society: reevaluating the deficit model of public attitudes. *Public Understanding Sci* **13**, 55–74.

Looking Forward to the Nano Situation

Lessons from the Bio-Decade: A Social Scientific Perspective

George Gaskell

Introduction

Recent times have seen a remarkable shift in the views about the nexus of science and society and its implications for the development of nanotechnologies. Consider the following quotations from influential reports on the future of nanotechnology. First, from the Royal Society and the Royal Academy of Engineering:

> In the near- to medium term, many of the social and ethical concerns that have been expressed in evidence are not unique to nanotechnologies.

What Can Nanotechnology Learn from Biotechnology?
ISBN: 978-012-373990-2

The fact that they are not necessarily unique does not make these concerns any less valid. Past experience with controversial technologies demonstrates that effort will need to be spent whenever significant social and ethical issues arise, irrespective of whether they are genuinely new to nanotechnologies or not. We recommend that the consideration of ethical and social implications of advanced technologies (such as nanotechnologies) should form part of the formal training of all research students and staff working in these areas. (Royal Society and the Royal Academy of Engineering, 2004)

And second from a High Level Expert Group of the European Commission:

A Societal Observatory of Converging Technologies: The expert group recommends the creation of a standing committee for real-time monitoring and assessment of international CT research. The primary mission of this observatory is to study social drivers, economic and social opportunities and effects, ethics and human rights dimensions. It also serves as a clearing house and platform for public debate. Among the core members of the committee should be social scientists and philosophers. (Nordmann, 2004)

I think it fair to say that the tenor of these comments, on the significance of ethical and social issues and the need for a social observatory, represent a culture change from the 1990s, and evidence institutional learning as a result of the years of controversy over biotechnology. And coming, as they do, from within the groves of the scientific establishment, they are all the more striking. This is not to say, of course, that such views are necessarily widely applauded among the scientific community at large. In Europe, for example, there have been grumblings about the requirement to address social and ethical issues in research grant applications in 6th Framework programme. Cultural change is a slow process even when some of the elite are in the driving seat.

However, this is not our current concern. Rather, in this chapter I want to explore some of the misapprehensions that appear to have guided the introduction of agricultural biotechnologies and also to identify some of the socio-psychological processes underlying public anxieties about aspects of modern biotechnology. The failure to heed the warning signals of concern from consumers, citizens, and civil society, sometimes dismissed as mere "irrationality," stalled European innovation and cost the agbiotech companies a small fortune. And, as the wider scientific community looked on, it provided a formative learning experience on science in society, which could and should have important implications for the socially robust development of nanotechnology.

With the advent of nuclear power, computers, and most recently modern biotechnology or the life sciences, the three strategic technologies of the post World War II decades, a cleavage between science, technology, and society opened up. Increasingly, sections of the European public questioned whether the good life, as defined by science and technology, is actually what they, the public, aspire to. This cleavage turned into open conflict in Europe over genetically modified (GM) crops and food; a controversy that became emblematic of the questioning of scientific expertise and of the established procedures of risk governance. What were the roots of this controversy and can those developing nanotechologies, the next strategic technology, avoid a similar debacle?

Understanding the process of innovation

In the 1990s, the foremost strategic technology was the life sciences project, a vision of a unified scientific and industrial enterprise embracing medical, pharmaceutical, industrial, and agri-food technologies based on recombinant DNA. The life sciences project was seen by industry, academia, and governments as a transformative technological innovation for the twenty-first century. Bayer, Novartis, and Monsanto, all multinational companies, were among the leading players in the project, of whom Monsanto became probably the most high profile, albeit not always in quite the way they might have hoped. Monsanto's corporate strategy could be used as a model for a management textbook. A persuasive vision to become a world leader in agricultural biotechnologies; good investor relations bringing capital to sustain a significant R&D program and an aggressive acquisitions policy; successful lobbying of the US government and regulatory agencies, and the cultivation of customer relations through close links to the US farming community and farmers in many other countries. Essentially they based their strategy on what might be called the market model of innovation, depicted in Figure 12.1. Here success in the process of innovation is contingent on gaining the support of the regulators and the market forces. Having achieved this,

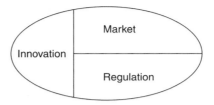

Figure 12.1   Market model of innovation

Monsanto was all set to become the global agbiotech company, the Microsoft of the life sciences.

In the event, the Monsanto's strategy floundered, the life sciences project collapsed and increasingly the "red" (medical) and "green" (agricultural) biotechnologies bifurcated. This was not due to a change in heart of the regulators or the market turning against GM crops. Rather, as Donald Rumsfeld put it "stuff happens." The stuff was a mixture of history and current events only some of which were directly related to biotechnology. Briefly, since the 1970s the European public had been uneasy and troubled by the idea of gene technology.

- A Eurobarometer survey in 1979 showed that 49% thought that synthetic food (what we now call GM food) was an unacceptable risk (Cantley, 1992).
- Then the 1990s saw a number of high-profile food crises in Europe—BSE, foot and mouth disease in cattle, and dioxin traces in chickens. In particular the BSE crisis evidenced the emergence of the risk society as the disease was clearly a consequence of human action and also one that did not respect national boundaries. BSE challenged the public's confidence in modern farming, scientific expertise, and the adequacy of the regulatory processes.

Into these troubled waters the first shipment of Monsanto's GM soya arrived in Europe in 1996, GM soya that was to enter the food chain without labeling. The rejection of labeling was a cardinal error—the denial of consumer rights nourished anti-globalization sentiments and without labeling any possible risk falls into the category of involuntary, for which there is dramatically lower risk tolerance than for voluntary risks (Starr, 1969).

A few months later in February 1997, the public announcement of "Dolly the Sheep" led to an explosion of media coverage, with

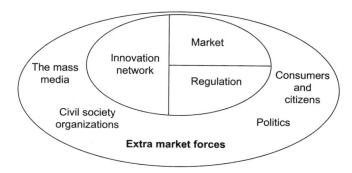

Figure 12.2 Societal model of innovation

discussions about the possibility of human cloning. For some this was a harbinger of what a biosociety had in store (Einsiedel *et al.*, 2001). Subsequently research was widely reported claiming evidence for environmental risks to the Monarch butterfly and health risks to rats. As sections of the European media and public became more polarized against GM agriculture, Monsanto embarked on a public relation charm offensive in Europe—it charmed few and offended many. Mounting public concern was accompanied by the disruption of GM field trials, supermarket boycotts of GM food, declarations of GM-free zones, and eventually a Europe wide de facto moratorium on the commercialization of GM crops across the European Union (Hampel *et al.*, 2006).

What was Monsanto's mistake? They ignored the warning signals and failed to appreciate that their innovation—GM food—was entering a more complex environment than the market model of innovation proposes. In the event the "Life Sciences Project" collapsed under the weight of unexpected resistance from the European public sphere—citizens, non-governmental organizations (NGOs), sections of the mass media, and political parties. And Robert Shapiro, Monsanto's CEO, admitted as such in an interview in 1999 (Kilman and Burton, 1999).

> We did proceed on the basis of our confidence in the technology and we saw our products as great boons both to farmers and to the environment. I guess we naively thought that the rest of the world would look at the information and come to the same conclusion.

In reality Monsanto was entering a more differentiated environment than they had envisaged (Figure 12.2). The societal model of innovation recognizes that success is not merely contingent on the reactions of those who are directly affected, but is also influenced by the wider public sphere. Here, we find a number of actors: the public as potential consumers (making purchasing decisions in the marketplace) and as citizens (exercising preferences in political contexts); NGOs operating in the space between the public, big business, and government—largely environmental and consumer groups—and organs of the mass media and politicians who saw the GM issue as an opportunity for increased circulation and political advantage respectively.

Furthermore these extra-market forces are not restricted to national boundaries. The globalization of industry and of risks has led to the creation of transnational issue fronts. Witness, for example, the rise of Greenpeace in many countries and the internationalization of debates on Dolly the sheep, and more recently on stem cell research.

The difficulty for innovators relying on the market model is that they may be unaware, or misperceive the extra-market forces and, as such, be unable to cope with these external challenges. Taking into consideration the changing value structures in modern Western societies, it may be expected that extra-market forces, seen particularly in the reactions of NGOs, the media, and the public as citizens rather than consumers will become an increasingly important contingency in technological innovation. The implication for the development of nanotechnology is that relevant environment needs to be mapped out ex ante, for if it is only recognized ex post it will be, in all probability, too late.

Questioning sound science

A focal issue in the GM food debacle was the confrontation between scientific and commonsense ways of thinking about risk and uncertainty. This emerged as a fundamental faultline in society and has implications for technological innovation in general. At a given time a society will face many different claims about potential hazards to public health and safety. For regulators and their scientific advisors the natural science paradigm provides the toolbox for determining which claims should be taken seriously and how to manage such hazards. The paradigm leads to a clear distinction between "relevant" hazards, those that may require action, and "irrelevant" hazards, those that do not. Those hazards defined as relevant are represented as risks. The criteria for the categorization of relevance are that the hazard can be understood in terms of current science and that it can, in principle, be quantified. In this way some of the range of claimed hazards are treated as familiar through the process of anchoring in contemporary scientific knowledge, and those that are unfamiliar are rejected from consideration as irrelevant.

With the label "risk" the familiar hazards are objectified, they become properties of the environment, that are further objectified in one of the most familiar frameworks, that is, numbers and the metric of probability.

The outcome of this anchoring of hazards in current scientific knowledge and objectification, first by the label risk and second with measurement in the form of numerical probabilities, reifies the relevant hazard into an "objective risk." Thus a risk refers to a real world state of affairs independent of perception or experience. Since this approach follows the canons of natural science, and natural science is seen to be rational and value free, so is the concept of objective risk also seen in these terms. As such it is attributed the status of "sound science" in regulatory discussions.

Deciding whether an objective and quantified risk is acceptable or not, the next step in the process of risk management, may be based on two quantitative indices in what is called evidence-based policymaking. First, a financial comparison of the benefits and costs weighted by probabilities, and second a comparative judgment against other known objective risks.

There are some important implications of the "sound science" approach. Sound science establishes the definition of what constitutes a relevant risk—those that have an established basis in scientific research—and at the same time bars from consideration other "risk claims." In this sense "sound science" determines the rules of evidence and becomes a filter of the "truth"—known risks are included but anything else is rejected from consideration as merely an hypothesis or even non-scientific fantasy. "Sound science" has another important implication in that it determines who is, and who is not, considered to be an expert. In other words whose voice should be heard in discussions of risk and safety and whose can and should be ignored. For those advocating sound science and "objective risks," the answer is almost a self-evident truth. Those who understand risk are the scientific experts, others who argue from different perspectives merely muddy the waters.

While faith in sound science's approach to risk identification gains some validity from the past record of successful risk management of many technologies, the approach is not without its critics. Some have argued that the framing or specification of what constitutes a risk is not free of value judgments and, furthermore, that the assessment of risks presupposes certain tractable and often highly debatable assumptions. Equally debatable are assessments of the putative benefits and costs. While a rational choice approach specifies a normative principle for weighing up risks, benefits and costs, what the GM debate illustrated very clearly is that, even accepting rational choice as a guiding tool, a key issue is who defines the risks, the benefits, and the costs? On both the nature and extent of these three attributes the promoters and critics of GM technologies had rather different ideas.

A controversial extension of sound science is seen in what is called functional similarity or functional equivalence (Kessler *et al.*, 1992). Here, an innovation that is unfamiliar to science is rendered familiar by anchoring it, or likening it, to something that is known to science. The following example outlines the reasoning; conventional tomatoes are not a public health risk and since GM tomatoes are functionally similar in all significant aspects, then they can be assumed to be equally safe. As such there is no need for regulatory oversight of the GM variety. Now, the question is whether the criteria for and decisions about functional similarity are based solely on science, which might be seen as unobjectionable, or whether lobbying or regulatory and economic expediency might inform such decisions. For the Food and Drug Administration in the United States, the concept of functional similarity was used, some claim, as part of a light touch approach to the regulation of GM crops and food (Jasanoff, 2005). By contrast, European legislation on GM, the outcome of a compromise between opponents and supporters of biotechnology, took a different, and far reaching, approach. Here, it was decided that since the process of production differed (conventional tomato versus rDNA tomato), then different regulatory arrangement were required. How, one might ask, will nano-products be judged regarding functional similarity, on what criteria, and with what consequences for the technology and the public's confidence in the regulatory arrangements?

Another extension of sound science is seen when the uncertainties are recognized by scientists to be beyond the frontiers of current knowledge. At this point, Sir Robert May, a former Chief Scientific Advisor to the UK Government, captured the essence of the scientific approach in the following quotation: "There are so many unknown factors and so much scientific ignorance, that top calibre advisers are needed to guide us through the fog" (Royal Society, 1999). Hence, even when the scientists know that they don't know, they still know better than anyone else. In this way science maintains its privileged position as the only legitimate authority on considerations of uncertainties related to impacts of a new technology.

In defence of sound science scientists and regulators might argue, "On what other basis can a society proceed?" Without a criterion of acceptable evidence and a basis for judging expertise, any claim from whatever source, including malicious sources, would have equal weight. And if this were the case then society could be at the mercy of Luddites and other fringe opinions. Increasingly, however, these views are challenged by a growing recognition that the constitution of societal risks is much more complex than strict scientific assessment. Defining the scope of the "relevant risks," and whether they are worthwhile, takes the issue into the public domain. As the European Commissioner for Research, Janez Potocnik, commented at a recent conference, "Of course there is the question is it safe? but, equally important is the question is this a world in which we want to live?"

Perspectives on risk

Thompson and Dean (1996) propose that current models of risk fall on a continuum between two extreme positions. What has been referred to as sound science is an exemplar of the probabilistic formulation. By contrast, contextualist formulations embrace a fuzzier definition of risk, allowing for the inclusion of characteristics that are unrelated to probabilistic assessment, for example social and cultural values. Cultural theory is the exemplar of the contextualist formulation. Douglas and Wildavsky (1982) offer an explanation as to why different social groups have different attitudes towards technological and natural dangers. In her earlier work, the anthropologist Mary Douglas claimed that in any given culture the content of beliefs about purity, danger, and taboo is essentially arbitrary. Within a particular culture these arbitrary beliefs become fixed and serve to organize and reinforce social relations according to hierarchies of power. Rayner (1992) argues that the social construction of risk occurs not only at the societal level but can also be observed within smaller collectives such as firms, political parties, and NGOs. The implication for the study of risk is important because it shifts the emphasis away from differences or biases in the perception of objective risks towards social processes and more fundamental types of intergroup cleavages.

Douglas and Wildavsky (1982) proposed four prototypical cultural types within modern industrialized societies. These are located along two dimensions that describe first the degree of social incorporation constituted within the culture and second the nature of these social interactions. While attempts to empirically corroborate this approach have met with limited success, it does resonate with the history of debates around biotechnology. It suggests that for different groups in society—in industry, government, and public interest organizations, the specific arguments for and against biotechnology arise from different, and in some cases incommensurate "world views."

Let us explore these world views with the help of Figure 12.3. For the entrepreneur (the industrialist and some scientists), nature is seen as bountiful, benign and malleable; biotechnology with its promise to improve on what nature has provided, is seen as a golden opportunity. In this world view the concept of risk is defined in terms of sound science. By contrast the egalitarians, exemplified by the "Greens" and Prince Charles, see nature as a delicate and precarious system and fear that any interventions may lead to unforeseen and potentially dire consequences. For them sound science is irrelevant, because it is the risks of biotechnology that are unknown and even unknowable that are of concern. Hence they argue for precaution.

Fatalist	**Bureaucrat**
The "cynic"	Regulators and government
Nature capricious	**Nature is tolerant within limits**
"In the lap of the gods"	Managing risks
Entrepreneur	**Egalitarian**
Industrial scientists	Prince Charles and the "Greens"
Nature benign	**Nature is fragile and venerated**
Sound science	Risks unknown
Objective risk	Precaution

Figure 12.3 Four stereotypes from "cultural theory"

This typology, while suggestive of more polarized positions, does not fully capture the structure of public opinion. The evidence of survey research and of qualitative studies we have conducted suggest that people are ambivalent. They view biotechnology through both the entrepreneurial and egalitarian perspectives. As entrepreneurs, people welcome progress and recognize the many contributions of science and technology to everyday life. But as egalitarians, they are deeply troubled by some applications of modern biotechnology which are seen to threaten the natural order (Wagner *et al.*, 2001).

For sections of the public, the idea of reducing the potential dangers of biotechnology to known risks, those that can be quantified, is not persuasive. In the public mind there are risks and uncertainties inherent in doing things that upset socially and culturally defined ways of acting, and to do such things is often described as immoral. In this sense the distinction between risks and ethics is meaningless for the public, the two concepts are intermingled.

While the public are aware of the possibility of side-effects from medicines and assume that these are the subject of scientific monitoring, when talking about the strange world of biotechnology, they entertain two other types of concerns. The first concern flows from the novel procedures used in biotechnology. These are widely perceived to be a threat to the natural order and to have unknown, perhaps unknowable consequences in the years to come. These we might call "moral hazards," in so far as to promote the technology today in the light of such possibilities, and without plans to conduct the necessary safety research appears to the public to be immoral. The second concern relates to the role of science and technology in society. This concern goes beyond biotechnology, but it seems as if biotechnology brought it to the fore. People worry about Where will technology lead to? Who is in charge? Are scientists independent of, or accountable to the biotechnology industry? Who should decide on the future shape of society? and Who is there to speak up for the interests of the ordinary person? These might be called "democratic hazards."

Returning to Figure 12.3, in an ideal world perhaps the bureaucrats (regulators and politicians) would mediate between the entrepreneurial and egalitarian positions, institutionalizing the ways in which a consensus is reached that serves the public interest. What many of the public feel is that biotechnology has been developed and regulated within the entrepreneurial world view at the expense of other values and conceptions of risk. In the belief that these other perspectives have been sidelined and ignored, people's confidence and trust in the technology and in the regulatory processes have been called into question.

Menacing images and magical thinking

For many people genetic modification is still a new and unfamiliar technology. Some may understand it through formal or informal education, others from media coverage, past experiences of technological innovations, and popular cultural images. Such connotative meanings may lead people to see genetic modification as an advance on traditional plant or animal breeding, or as something akin to Frankenstein or Jurassic Park. From conversations with groups of the general public we developed three survey questions addressing what we called "menacing images." The survey asked whether people agreed or not with the following statements:

- Ordinary tomatoes don't have genes but genetically modified ones do.
- By eating a genetically modified fruit, a person's genes could also become modified.
- Genetically modified animals are always bigger than ordinary ones.

While agreement indicates an absence of knowledge about genetics, it also attests to the idea that food biotechnology is associated with primordial fears of respectively adulteration, infection, and monstrosities. Between 20 and 30% of Europeans agree with these propositions and while the percentage of people giving the correct answer has increased by on average 7% between 1996 and 2005, they appear to tap into beliefs that are resistant to change (Gaskell *et al.*, 2007).

With these questions we had stumbled on the work of various nineteenth-century anthropologists, including Sir James Frazer (1930), on sympathetic magic, recently taken up by Rozin and Nemeroff (1990). It may be 200 years after the Enlightenment, but traditional ways of thinking still inform the ways in which people make sense of some current events, particularly in relation to food and disease. To take a recent example: as the bird flu epidemic in

Asia was featured in the Western European media, the household consumption of chicken declined dramatically. Here the scientific evidence is unequivocal—people cannot catch bird flu from the consumption of cooked chicken—but apparently unconvincing. This is an example of the law of contagion, the transfer of essences and summarized by "once in contact, always in contact." Genetic modification appears to invoke these "primitive" ways of thinking, science may claim to have the rational answer, but magical thinking can sometimes trump rational argument.

That about 25–35% of Europeans assent to these menacing image propositions does not necessarily mean that they all held such views before being asked the question in the interview. In all probability many would not have come across the issue before. But when the question is posed, people try to make sense of it, as best they can. Perhaps a combination of their unease about the technology, anxieties about food, and magical thinking lead them to assume the worst.

Uncertainty and anxiety

As noted by Robert May, it is recognized that developments in biotechnology enter areas of scientific uncertainty necessitating further research. But, how does the public respond to uncertainty? For suggestive evidence we turn to Daniel Ellsberg, a sometime victim of the Watergate plumbers. Ellsberg (1961) offered people (mainly economists as it happened) a choice between two proposed bets based on the flip of a coin. In the first proposition an unbiased coin is used, thus the probability of heads is 0.5, as is the probability of tails. In a single flip of the coin, if it comes up heads you win $100, and if it comes up tails you get nothing—let's call this bet A. The second proposition also offers the opportunity to win $100 or getting nothing based on a single flip of a coin. But here, and crucially, the probability of heads is created by a random number generator—heads could have any probability between 1.0 and 0.0. Of course the combined probability of heads and tails adds to 1.0—let's call this bet B.

Now, if you were offered a choice between these two bets to win $100, which would you prefer? The so-called Ellsberg paradox is that the majority of people have a confident preference for bet A. It is a paradox because the result runs counter to rational choice. The expected value of both bet A and bet B are identical at $50. In bet A there is a 0.5 chance of winning $100, hence an expected value of $50 and the same holds for bet B.

According to the canons of rational choice people should be indifferent between the two bets, but the majority are not, why? Ellsberg suggested that the reason why people prefer bet A over bet B is that the latter is ambiguous. I would prefer to say that bet B appears to be more uncertain than bet A and for most people uncertainty is unattractive and, if possible, to be avoided. Indeed other research has shown that people will actually pay more to avoid situations where unknown probabilities are involved (Becker and Brownson, 1964).

The implication for technological innovation is that uncertainties make people uneasy. When scientists and specialists say of a technology "on present evidence we think it is safe, but there may be uncertainties in the future" or that "any possible risks will be sorted out if and when they are apparent," I suspect that the public is not greatly reassured, unless as we will see in the next section, there are substantial benefits to outweigh possible risks. Given the uncertainties about the toxicity of nanoparticles (echoes of asbestos?) and their current and prospective use in cosmetics, food, and household products, one wonders whether the public will be troubled when they learn about it?

More generally, if it is the case that uncertainty is becoming more pervasive (although it may be that societies are just more sensitive about it), we need to find better ways to discuss the issue and to arrive at socially acceptable thresholds and management strategies.

Weighing up gains and losses

Behavioral decision theory introduced rational choice models, the idea of decision-taking on the basis of maximizing utility, into psychology. However, anomalous findings from empirical research raised some objections to this formulation. In response, Kahneman and Tversky's "prospect theory" elaborated a general framework for understanding why people's actual behavior, in relation to risky decision-making, departs from the predictions of rational choice theory (Kahneman and Tversky, 1979). Prospect theory includes weighting functions for both probabilities and utilities. The probability function captures the findings on systematic biases of estimates of fatalities. People tend to overestimate (weight) low-probability events and underestimate those with a high probability, essentially a regression effect. Although the availability or vividness bias is one possible explanation, in prospect theory it is proposed that the over-weighting of low-probability events occurs regardless. That something is conceivable appears to be sufficient to give it a reality beyond its objective probability. The implications for new technologies are that even a hint of potential problems may loom significantly in the public mind.

The value function is defined in terms of gains and losses from a reference point or adaptation level. For gains, the function is con-cave and while the same holds for losses, in this context the slope of the curve is much steeper. Consider, for example the pleasure of the first cup of coffee in the morning—it is just what is needed to get going. But, by the time one has consumed three cups, the shakes set in and one is beyond diminishing marginal utility. By contrast, if one is eagerly anticipating the first cup, the pain of finding that your partner has drunk all the coffee is far in excess of the pleasure that the cup would have brought. In more technical terms the utility weighting leads to an asymmetry between "objectively" equivalent gains and losses (Figure 12.4).

Thus the pain from a small loss from one's current position will far outweigh the pleasure from an equivalent small gain. In terms of the way people think about biotechnology or nanotechnology, it

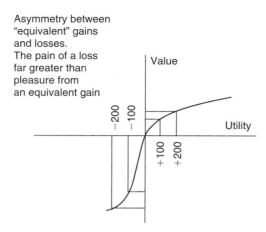

Figure 12.4 Prospect theory: weighing up gains and losses

follows from prospect theory that the potential harms might loom larger for the public even if weighed against "equivalent" (in terms of its formal expected value) gains in efficiency, reduction in price, or whatever. In other words, the benefits of any innovation need to be great in order to justify taking any risks.

Further research provides some interesting insights into the key role of benefits in the formation of public opinion. A sine qua non of an innovation is that it offers benefits over and above what is currently available. An innovation that offers no added value is almost an oxymoron. Benefits may be seen in lower costs, more functionality, enhanced quality, or even more symbolic value. The industry claimed GM crops would bring a range of benefits, including higher productivity and lower pesticide costs for producers; less environmental pollution from pesticides and herbicides, and new crop varieties to ameliorate hunger in developing countries. However, the first generation of GM commodity crops offered no direct benefits for consumers, a point that Robert Shapiro, Monsanto's CEO, lamented in an interview (Kilman and Burton, 1999): "Certainly if our first products had been something that had health benefits, it would have been easier to make our case," he said, "Then many consumers would have seen benefits, like has happened in the pharmaceutical world, where biotechnology isn't such a cause of opposition".

This absence of benefits for the public was the Achilles' heel of agricultural biotechnologies. In the spirit of utilitarianism, people said words to the effect "Our food is pretty good already, is GM necessary? Are there any benefits for me?" And with "no" as the answer to the final question, what rational person would wish to support such a non-innovation, particularly when there were suspicions about risks? And as the second generation of GM crops with promises of consumer benefits has yet to arrive, there is little to persuade the doubters that this really is an innovation.

Did the absence of benefits accentuate risk perception? While our research suggests that this is not the case, it clearly illustrates the problem of promoting a technology without a value-added component. In an analysis of the 2002 Eurobarometer survey on biotechnology we identified four different groups of respondents based on a classification of their risk and benefit perceptions regarding GM food (Gaskell *et al.*, 2004). The "relaxed group" perceive benefits with no risk; the "trade-off group" perceive both risks and benefits; the "skeptical group" perceive risks and no benefits and a rather small group, the "uninterested" see neither benefits nor risks.

Fully 60% of Europeans fell into the skeptical group in relation to GM food. Using multivariate analyses we modeled the decision-taking strategies of the "trade-off", "relaxed" and "skeptical" groups and showed that they differed in respect of key social and cognitive resources that may inform their views of GM foods.

As people perceive greater levels of benefit, so does risk perception increasingly enter into their judgments of support for GM food. Conversely, as perceived benefit declines, so is the effect of risk perception on support attenuated.

In other words, if benefit is perceived then the respondent goes on to think about risk and these two attributes are combined into an overall judgment of encouragement (the strategy of the "trade-off" group). By contrast, for the "skeptical" group, the absence of perceived benefits acts to truncate their deliberation on the issue; the attribute of risk is deemed irrelevant and accordingly has less influence on the final judgment of support. Here, the implied decision model is lexicographic, possibly based on Slovic's affect heuristic. One attribute, the absence of benefit, is dominant.

For the "relaxed" group the decision-making strategy is far from clear. Their perception of benefits may lead them to ignore the risks (lexicographic) or they may deliberate on the risks, judge them to be minimal and combine the two attributes according to the rational choice model.

These analyses lead to a paradoxical conclusion. Risk only appears to matter when people recognize that benefits are on offer. For innovators, the obvious implication of these analyses is to highlight the benefits of any new development. However, there is but a small step between a "highlight" and public relations hyperbole. Thus, for example, claims that GM crops and food would ameliorate food shortages in the developing countries were simply not believed by many Europeans. While there are commercial and other pressures to make bold claims about potential impacts of scientific research and technological innovations, these need to be credible.

Truth claims and communicating science

Whether it is explaining risk and uncertainty to journalists or debating aspects of a new technology in the media with, for example, environmental pressure groups, scientists often feel ill at ease. What they take as a valid argument follows the logic of the scientific article. Science is about making claims warranted by the empirical evidence that meets the criteria of approval via the peer review process. If there is no concrete evidence scientists are often reluctant to speculate. However, their critics are sometimes less constrained by "the evidence" and indeed there have been carefully orchestrated campaigns designed to heighten risk awareness of, for example, GM food and nanotechnology. Activists opposing GM crops have donned chemical warfare protective clothing as they tore up fields of GM maize, and ensured that a television production team is on hand to film the proceedings and relay them to the public at peak times. Why do such exhibitions appear to provide better evidence than reports of the scientists' careful work in the laboratories?

Bruner (1985, 1986) provides the clue. According to his work, there are two modes of thought that capture the way people represent and explain events in the world, what is taken to be "true:" *Paradigmatic thought* is essentially the rational scientific approach, characterized by abstraction and generality. It is context-free, timeless, and universal. It seeks to establish truth through verification procedures and empirical proof. In this mode, action is understood in terms of causes and correlations. By contrast, the *narrative mode of thought* is based on story-telling. It does not seek to establish truth but truth-likeness or verisimilitude. It is not timeless but temporal, and it sees action as a result of human intentions. To this end, it constructs two landscapes simultaneously: that of action and that of intention. A good story recounts not only what happened and who did what, but also why they did so. The narrative form is part of culture's toolkit (Swindler, 1986), which provides people not only with a set of stories and myths, but also a generative structure for the composition of a good account. A given culture has a set of symbols available for sense-making and explaining actions and events. A good story will use them to gain acceptance and believability. To this extent, the narrative mode of thought operates within a horizon of familiar cultural symbols.

Bruner (1985, 1986) argues that it is not possible to understand the operation of either the paradigmatic or the narrative mode of thought in terms of the other. While the imaginative application of the paradigmatic mode results in "good theory, tight analysis, logical proof and empirical discovery guided by reasoned hypothesis," the imaginative application of the narrative mode leads instead to "good stories, gripping drama and believable historical accounts." To this extent the narrative mode is not a primitive or vulgar form of paradigmatic thought. In the search for what constitutes the truth in context, the two modes are based on different versions of reality and conceptions of evidence, valid in different contexts and for different social groupings.

If Bruner is right this presents a dilemma for science communication. For communication within the scientific community the paradigmatic mode is appropriate, but for communication with other audiences it may not be persuasive, particularly when it is challenged by a narrative account. In the UK this has been seen in a long running dispute over the MMR (measles, mumps, and rubella) vaccination for children. The epidemiological evidence (paradigmatic) is overwhelmingly in support of the efficacy of the triple vaccine. Yet a single paper in a medical journal, the *Lancet*, based on a handful of cases, purported to show a link between the triple vaccine and childhood autism. The paper attracted considerable media attention, including interviews with distressed parents of autistic children claiming that the disease was due to the triple vaccination. This led to a significant fall-off in the rates of vaccination and opened up the possibility of epidemics of dangerous childhood disease.

Confronted by such situations, laments are still to be heard about the quality of science education in schools and the need for a scientifically literate public. Without wishing to gainsay the value of education as a societal good, the "educating the public" strategy to alleviate scientific controversies has met with little success; but it also misses the point. New thinking argues that what is needed is a more inclusive debate about science, technology, and society. A debate that is an exchange of views rather than a one-way flow of information from the experts to the public.

This will require scientists and policymakers to recognize and listen to the "truths" and concerns articulated in narrative thinking. Rather than seeing the public as the problem and in need of education, maybe the locus of the problem and its solution are with the scientific community itself. If they want the public have confidence in and on occasions to participate in discussions about technological innovation, then perhaps it is up to them to bridge the gap between paradigmatic and narrative thinking. And to achieve this, they would need to learn to speak in the narrative mode and to engage with the public in terms that will be understood.

Changing science, changing societies

Was the GM controversy a unique case that will soon be forgotten? Or is it an indication of wider changes in society and a portent of problems for future areas of technological innovation, for example developments in genomics, nanotechnology and the so-called converging technologies? I think that the evidence points to a sea change that cannot be ignored.

Currently we are seeing changes in both science and society (Gaskell and Bauer, 2001). Science is evidently more commercialized, less accountable to the peer group of science, and more accountable to the financial markets; and this at a time when many new developments, such as genetic testing and stem cell cloning raise many scientific, ethical, and moral uncertainties. Such innovations raise questions that often go far beyond the coping capacity of traditional regulatory frameworks.

In parallel we see the emergence of a better educated but more skeptical society. Deference to and trust in the traditional hierarchies is declining as people ask "Why should I trust these experts? Do they understand my values and concerns? Are they acting in the interests of the wider public?" The public are prepared to trust, not those who demand it, but those who show, by their actions, that it is merited.

Increasingly, controversies over science and technology are becoming more political, reflecting multiple rationalities. Fundamental questions, such as What is a desirable future for society and how can science and technology contribute to this? Who should bear the risks of new technologies and who should decide on these matters?, are more to do with values than the esoterics of science.

The adoption of public consultation and participation is, in part, a recognition of these societal trends. Interestingly, when the public is consulted on technological innovation, they are rarely anti-technological Luddites. Looking back at GM consultations in the early 1990s, we find that the public's voiced concerns are now embodied in legislation.

However, there is a need to clarify the objectives of public participation, the types of issues where it would be appropriate and how it relates to regulation. My view is that it should be seen as part of the process of socially robust technological innovation. Participation has a role to play when values are at stake and where risks are disputed. It should be seen as a complement to effective regulation, not a substitute for it.

The public will support innovation when it is compatible with social values, and be prepared to accept risks in the context of real benefits. Given the opportunity to deliberate on such innovations, the public voice can be expected to be measured and moderate. The challenge is to ensure that society is well represented in the development of science and technology and equally that science is well represented in society.

Implications for nanotechnology

What are the lessons from agricultural biotechnology for nanotechnology and other technological innovations?

- Proactive engagement with the public sphere is needed.
- Avoid involuntary exposure to nano-products; labeling is not merely a consumer right but also prudent commercial policy.
- Do not allow "sound science" to trump social values.
- Ensure the products of nanotechnology have tangible consumer benefits.
- Learn to speak to the public in terms that they can understand.
- Avoid scientific hubris. Anyone believing that nanotechnology will solve the world's problems needs a lesson in the history of technology.
- Recognize that society is changing, deference is a scarce commodity, and trust has to be earned.
- For the public, progress is valued but only so long as it resonates with social values.
- Don't re-invent the wheel. The social sciences and humanities have much to offer the process of socially robust technological innovation.

References

Becker, J. W. and Brownson, F. O. (1964). What price ambiguity? Or the role of ambiguity in decision making. *J Political Econ* **72**, 62–73.

Bruner, J. (1985). Narrative and paradigmatic modes of thought. In: Eisner, E., ed. *Learning and Teaching the Ways of Knowing.* University of Chicago Press.

Bruner, J. (1986). *Actual Minds, Possible Worlds*. Harvard University Press.

Cantley, M. (1992). Public perception, public policy, the public interest and public information: the evolution of policy for biotechnology in the European Community, 1982–92. In: Durant, J., ed. *Biotechnology in Public*. Science Museum.

Douglas, M. and Wildavsky, A. (1982). *Risk and Culture. An Essay on the Selection of Technological and Environmental Danger*. University of California Press.

Einsiedel, E. *et al.* (2001). Brave new sheep—the clone named Dolly. In: Bauer, M. W. and Gaskell, G., eds. *Biotechnology: The Making of a Global Controversy*. Cambridge University Press, 313–347.

Ellsberg, D. (1961). Risk, ambiguity, and the Savage axioms. *Q J Econ* **75**, 643–699.

Frazer, J. G. (1930). *The Golden Bough: Studies in Magic and Religion*. Macmillan.

Gaskell, G. and Bauer, M. W., ed. (2001). *Biotechnology 1996–2000: the Years of Controversy*. Science Museum.

Gaskell, G., Allum, N., Wagner, W., Kronberger, N., Torgersen, H., and Bardes, J. (2004). GM foods and the misperception of risk perception. *Risk Anal* **24**, 183–192.

Gaskell, G. *et al.* (2007). Europeans and Biotechnology: Patterns and Trends, Directorate-General for Research European Commission: Brussels.

Hampel, J. *et al.* (2006). Public mobilization and policy consequences. In: Gaskell, G. and Bauer, M. W., eds. *Genomics and Society: Legal, Ethical and Social Dimensions*. Earthscan, 77–94.

Jasanoff, S. (2005). *Designs on Nature: Science and Democracy in Europe and the United States*. Princeton University Press.

Kahneman, D. and Tversky, A. (1979). Prospect theory: analysis of decision under risk. *Econometrica* **47**, 263–291.

Kessler, D. A. *et al.* (1992). The safety of foods developed by biotechnology. *Science* June 26; **256**(5065), 1747–1749, 1832.

Kilman, S. and Burton, T. M. (1999). Monsanto Boss's Vision Now Confronts Reality. *Wall Street Journal*, December 21.

Nordmann, A. (2004). *Converging Technologies—Shaping the Future of European Societies*. Report of a High Level Group. European Commission.

Rayner, S. (1992). Cultural theory and risk analysis. In: Krimsky, S. and Golding, D., eds. *Social Theories of Risk*. Praeger.

Royal Society (1999). *Science, Technology and Social Responsibility.* The Royal Society.

Royal Society and the Royal Academy of Engineering (2004). *Nanoscience and Nanotechnologies: Opportunities and Uncertainties.* The Royal Society (UK) and the Royal Academy of Engineering.

Rozin, P. and Nemeroff, C. (1990). The laws of sympathetic magic. In: Stigler, J., Sweder, R., and Herdt, G., eds. *Cultural Psychology— Essays on Comparative Human Development.* Cambridge University Press, pp. 205–232.

Starr, C. (1969). Social benefit versus technological risk. *Science* **165**, 1232–1238.

Swindler, A. (1986). Culture in action: Symbols and strategies. *Am Sociol Rev* **51**, 273–286.

Thompson, P. and Dean, W. (1996) Competing conceptions of risk. *Risk: Health, Safety and Environment*, 361–384[Fall].

Wagner, W., Kronberger, N., Gaskell, G. *et al.* (2001). Nature in disorder: the troubled public of biotechnology. In: Gaskell, G. and Bauer, M. W., eds. *Biotechnology 1996–2000: the Years of Controversy.* Science Museum.

What Can Nanotechnology Learn from Biotechnology?

Lawrence Busch and John R. Lloyd

Introduction

Subsequent to the introduction of all new technologies, some emerge as winners and others fall by the wayside as losers. Consider, for example, the comparison between the new pharmaceutical biotechnologies and the agricultural biotechnologies. The new generation of pharmaceutical products is embraced worldwide, but resistance to agricultural biotechnology can be found almost everywhere, not least in the United States. Herbicide-tolerant and insect-resistant crops (mainly corn, soy, cotton, and canola) are now commonplace in several nations around the world. On the other hand, some attempts to introduce genetically modified (GM) crops have been almost completely blocked, perhaps permanently, through the efforts of both public opinion advocacy organizations and various non-governmental organizations (NGOs).

The message here is straightforward: Although agricultural biotechnologies have enjoyed a few successes, overall they have had more than their share of failures. Without doubt, they failed to live up to the claims of the early 1980s, but even discounting those claims as mere hyperbole, they have hardly done well. In point of fact, the agricultural biotechnologies have been the subject of continuing controversy almost since their inception. They have produced endless street demonstrations and protests, field burnings, debates, and even an occasional guerrilla action. They have even been panned by respected members of the financial community.

In contrast, the pharmaceutical biotechnologies have been warmly embraced by the public. Consumer electronics, personal computers, mobile phones, and the Internet have for the most part also been enthusiastically adopted by the public. What are the reasons for the differences in how the technologies have been received?

One might argue that there is something unique about the agricultural biotechnologies that led to the current situation. Yet, this appears to be unlikely. The modification of plants through conventional breeding never brought much public concern. Nor can we say that the controversy is the result of differences in food cultures. There is a gross exaggeration that suggests that Americans do not care what they eat while Europeans are quite fussy. But within Europe, acceptance of agricultural biotechnology is high in both the Netherlands and Italy, and low in both France and Norway. Nor can we ascribe the problems to the lead in this field taken by American companies such as Monsanto, Dow, and Dupont. European companies have been equally active in pursuit of new products with Syngenta (Switzerland) and Bayer (Germany) in the lead. In contrast, we argue here that the successful introduction of new products requires not merely that there be an invention and enthusiastic adopters, but that all the other actors, both within and with oversight over the supply chain, must be enrolled.

Will the new nanotechnologies encounter the same or similar resistance? Are there lessons that we can learn from by examining the failures and successes of agricultural biotechnologies? Can we shape the new nanotechnologies as well as respond to the concerns of critics and skeptics? Simply stated, what lessons can we learn from the experiences with the agricultural biotechnologies that will help us avoid the same result with the design of nanotechnological products and processes? What actions on the part of companies and governments might ensure the rapid and satisfactory resolution of concerns about nanotechnologies? What actions are likely to enhance public support for the promises that these new technologies bring? And what actions are likely to diminish that support?

In this chapter we begin by discussing the nature of scientific innovations. Then, we discuss the development of new technologies based on that science, specifically using the cases of Roundup Ready and Bt seeds to illustrate our point. The purveyors of nanotechnological products and processes can learn a great deal from a careful examination of these successful and failed examples of agricultural biotechnologies.

Scientific innovation

Scientific innovation begins in the pure science research laboratory and ends with the consumer. In between technologies and products are developed, manufactured, and marketed. Basic research is often the source of innovation of new technologies. Unfettered ideating and subsequent exploration without focus on downstream technologies that might develop from the process of discovery are fundamental to the identification of many contemporary new technologies.

Here we state that the identification of possible products—using the new knowledge base from basic and fundamental research—is where societal concerns begin to develop and emerge. How we handle the development of new technologies from this state forward is what controls acceptance or rejection of the new products and processes. This is where social acceptance is initiated or destroyed. We will use the new nanotechnologies as an example here.

As shown in Figure 13.1, there are five basic stages in the process of successful product innovation. The process begins with pure science and the generation of new knowledge. As the technical community begins to understand the new knowledge more thoroughly, technology developers, engineers, begin to convert that knowledge into useable forms such as concepts for new products or processes.

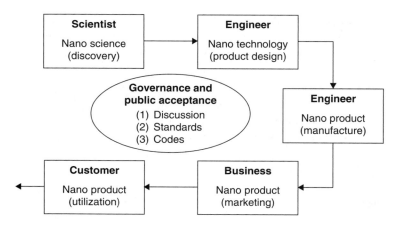

Figure 13.1 The product innovation chain

It is at this stage that a more universal interest by the public is generated in the fundamental knowledge that the scientist developed. We begin to find citizen groups and regulatory bodies exhibiting more and more "interest" in the new technology.

As new products are innovated, concerns for human, and now environmental, well-being become important considerations in the product design, to both the engineer and to the public. With the product designed and having moved through the prototype stage of development, the manufacturing community dominates the innovation process. New supply chain needs are identified, and this spreads the impact of the original product concept to a much wider community, even an international community.

With the ability to manufacture the new products, the marketing function begins in earnest. Gaining public acceptance is critical to the long-term success of the new technology. Concerns the impact of the innovation and product realization processes on the health of the consumer as well as the well-being of the environment will often become more important than the benefits brought to humankind by the new technology.

The consumer is the endpoint of the innovation process. The technology may or may not proceed as far as the consumer, but if it does, the consumer has the final input on determining success or failure.

As we will argue later, the impact of standards and regulation (governance in Figure 13.1) can be a powerful force in the acceptance process. With the safety concerns regarding the manufacture, utilization, and disposal of the new technology properly addressed, and pricing determined to be acceptable, from the point of view of the public and consumer organizations, through the establishment of sound standards and subsequent regulations, new and emerging technologies will find their way into our lives. Without that, they lose. Plain and simple, they lose. At what point should we begin the process of establishing standards or more stringent regulations in order to make the products and processes acceptable to the public? This is perhaps the most important question we can ask. We can seek some answers to this by examining as a case study the biotechnology field.

The process of innovation of new products in biotechnology

Diffusion theory suggests that innovations move through society much like objects in a vacuum (cf. Latour, 1987; Rogers, 1995). According to the diffusion theory, innovations encounter no friction, no resistance, and no stumbling blocks. Moreover, diffusion theory has tended to ignore the "networked" character of most radical innovations. Such innovations often require a range of social, economic, and technical changes before they can be effectively put to use.

Consider, for example, the Universal Product Code (UPC). It may be argued that the UPC transformed the retail grocery industry by reducing the time necessary for checkout and by improving inventory control, but it only did so by virtue of a host of ancillary activities and innovations throughout the supply chain. Manufacturers had to be convinced to print them on the labels of most packaged goods. Supermarkets had to buy scanners, and had to have sufficiently powerful computers to which to attach them. The industry as a whole had to establish the Uniform Code Council (and the European Article Numbering Association in Europe). Consumers had to be convinced that the absence of price labels on products did not mean that supermarkets were cheating them or spying on them. The Federal Trade Commission and other government regulators had to be convinced that the innovations were not a form of illegal restraint of trade. And unions had to be convinced that the new technologies would not require wholesale layoffs (Brown, 1997; Dunlop, 2001; Haberman, 2001). If any of these actors in the supply chain had failed to support the new technologies and organizational forms surrounding UPCs, we would still be ringing up groceries by pushing buttons on a cash register.

Similarly, agricultural biotechnology innovations have had to satisfy all the actors in the supply chain. To date, this has been only partially achieved, and thus complete acceptance has also not been achieved. That makes the study of the history of these innovations/technologies a good source of insight for what to avoid as today we launch efforts to bring nanotechnologies to market.

Processes of variable regulation of biotechnology

The Reagan administration was a staunch proponent of deregulation. Both Reagan's supporters and pure scientific researchers in the technical community argued that regulation stifled innovation, slowed the pace of technology adoption, and constrained market growth. It was during these years that the first products of plant biotechnological research reached the marketplace.

Monsanto executives, who were pursuing the conversion of the basic plant biotechnical research into new products, requested and received an audience with then Vice President, George Bush. They insisted that the Reagan administration must design procedures to regulate the new agricultural biotechnologies (Eichenwald, 2001). The initial reaction of the administration was unsurprisingly negative, but Monsanto's reasoning was unusually persuasive. The new plant biotechnologies were a subject of concern among many in the farming, environmental, and consumer communities. Regulation would quiet their fears by declaring, because of the imposition of regulations, that the new technologies were safe for both human health and the environment. Regulation had the power to enhance public confidence in the new products. Furthermore, Monsanto surely understood that regulation would not only reduce biotechnology companies' liabilities (were adverse affects to arise) that regulation would weed out the weaker companies, who would be unable to afford the costs of meeting the requirements of the regulatory process.

It appears that Monsanto and the Reagan administration soon agreed on the value of regulation. However, given the strong desire of the administration to limit regulation, it was decided that existing laws would be adapted to this new task. Thus, a "coordinated framework" of regulation was cobbled together from enabling legislation from the

Environmental Protection Agency, the Food and Drug Administration (FDA), and the US Department of Agriculture (USDA).

As Kurt Eichenwald (2001, A1) put it in the *New York Times*:

> It was an outcome that would be repeated, again and again, through three administrations. What Monsanto wished for from Washington, Monsanto—and, by extension, the biotechnology industry—got. If the company's strategy demanded regulations, rules favored by the industry were adopted. And when the company abruptly decided that it needed to throw off the regulations and speed its foods to market, the White House quickly ushered through an unusually generous policy of self-policing.

Furor over bovine growth hormone

Bovine growth hormone, later called bovine somatotropin, was the first biotechnology product to reach the market. It immediately created a public furor. As Monsanto CEO Richard Mahoney noted:

> We got into BST like we got into a lot of things. We'd been making agricultural chemicals for years. You increase the productivity of the farmer; you keep half [of the profits] and give him half. So what's the big deal? There wasn't even one discussion of the social implications. I never thought of it. (quoted in Charles, 2001)

Prior to 1990 Monsanto attempted to engage the critics. But things changed dramatically when the charismatic Robert Shapiro was appointed CEO. With great fervor, Shapiro argued to his staff and to the world that Monsanto was about to change the world for the better, and in the process would, not incidentally, amass a considerable fortune as well. Monsanto continued to lobby what was by then the first Bush administration, and, by 1992, Monsanto had managed to speed the regulatory process. Both the administration and industry leaders believed that science had shown the complete safety of the products of biotechnology. A few scientists at the FDA had raised concerns, but they were a distinct minority and could be easily ignored by both government and industry. Furthermore, the FDA ruled that it would not be necessary to label the new products unless they were significantly different from products already on the market. To do otherwise, it was felt, would create public concern where there was none.[1]

But as former CEO of Pioneer Seeds, Thomas Urban, argued:

> Monsanto forgot who their client was …. If they had realized their client was the final consumer they should have embraced labeling.

They should have said, "We're for it." They should have said, "We insist that food be labeled." They should have said, "I'm the consumer's friend here." There was some risk. But the risk was a hell of a lot less. (quoted in Eichenwald, 2001)

Instead of quelling public concern, the lack regulation of labeling raised concerns among a wide range of public interest groups representing consumer, environmental, small farmer, and animal welfare groups. By the end of the decade, those groups distrusted those pushing the technologies, and they had received sufficient support to challenge both the non-functioning regulatory regime and the technologies themselves.

But the agricultural biotechnology industry continued to apply pressure not to support labeling. Backed by industry, US trade negotiators pressured the Europeans. They argued that European concerns about environmental safety and human health were unfounded. In Europe, public protests erupted soon after.

The major actors

Scientists and engineers

The products were innovated and developed by the scientists and engineers. As the marketing arm of Monsanto began to try to sell the seeds to produce the new crops, the complexion of the response to the new technology began to evolve. The technology was, as we shall see, brought to this point with no feedback from the producers and consumers. The farmer was the key to the future of the product.

Seed companies

Seed companies have been easy to enroll in the biotechnology industry, but the path to enrollment has not been easy. Seed companies are the front line in dealing directly with farmers, who are often also their neighbors. Moreover, they have had to inform farmers that they would be required to sign a complex contract to "license" seeds, and that they would no longer be able to save seed as they had done in the past.

But it should also be remembered that, until recently, the seed market was made up of a few very large players (mainly in hybrid corn and vegetables) and a large number of very small "mom and pop" businesses. Overall, the business was stagnant, with little change in sales from year to year. The larger companies could, and sometimes did, build market share at the expense of competitors, but this avenue was only open to those with considerable capital. In contrast, GM crops offered seed companies an immediate gain in sales volume of 20–30% as farmers who could no longer replant would have to return to seed dealers to buy that seed. Moreover, when farmers complained, the seed companies could always argue that the idea to prohibit seed saving was not theirs, but came directly from the biotechnology companies. Monsanto's initial insistence on charging a separate "technology fee," supported the seed companies that made that argument (Charles, 2001), and imparted a bad impression of the new technology.

Farmers

If Monsanto and other large players in the agricultural biotechnology industry were not particularly concerned about final consumers, they were beginning to pay attention to farmers. Two newly modified crops were soon ready for market: those tolerant to Monsanto's herbicide Roundup (glyphosate), and those modified to enhance insect resistance through the addition of genes from *Bacillus thuringiensis*, a common bacterium often used as a spray by organic farmers. Unlike conventional seeds, the modified seeds did not increase yields significantly,[2] and in some cases even resulted in yield decline. But both Roundup Ready and Bt crops made farm management easier. The former allowed farmers to spray for weeds throughout the crop cycle, while the latter permitted a reduction in insecticide use. Not surprisingly, farmers embraced the crops lowering their costs and reducing the time devoted to management.

Monsanto also worked hard to prevent farmers from violating the terms of the licensing agreements, terms that prohibited replanting and resale of the crops. They established toll-free hotlines where informants could anonymously report violators to Monsanto. They also unsuccessfully lobbied state legislatures to require licensing of seed cleaners (Charles, 2001).

Despite the general acceptance among farmers, organic farmers spurned the GM crops. Given that USDA standards prohibit genetic modification, such farmers were and remained concerned that widespread use of GM seed would diminish or even eliminate their market niches. Moreover, given that pollen drifts from one field to another, organic and conventional farmers found themselves in what some described as a life and death battle against each other.

In sum, (most) farmers and seed companies were relatively easily enrolled by the biotechnology industry. In addition, the industry was highly successful in gleaning the support of the US Government, despite its relatively small size.[3] But the successes upstream in the supply chain were not easily repeated downstream. Indeed, there the industry continues to confront skeptical companies.

Manufacturers

Manufacturers have been wary of GM food products, fearing that some significant percentage of consumers would reject them. Arguably, the most dramatic rejection occurred with respect to potatoes designed to resist the ravages of the Colorado potato beetle, a major insect pest. Soon after the product was released Frito-Lay, the largest producer of potato chips in the US, and McDonalds, the largest user of french fries, both announced their intention of avoiding the modified crop (*Nation's Restaurant News*, 2001; Pollack, 2001). This decimated the market, leading Monsanto to withdraw the product from the market.

GM flax encountered a similar fate in the Canadian market. European buyers, purchasers of 60% of the crop, announced their intention to avoid the modified flax. Moreover, its rather strange name—Triffid—the name of a plant in a science fiction story of the 1950s that rendered those close to it blind, hardly exuded confidence. GM sugar beets suffered a similar fate when Mars and Hershey, major customers for sugar refiners, decided to avoid use of the product (Kilman, 2001). More recently, Monsanto abandoned plans to introduce GM wheat, perhaps due to resistance from bakers and millers, as well as from farmers concerned about the export market. Indeed, one observer reports that it is relatively easy for importers to get non-GM wheat from other nations where no costly segregation would be necessary (Wisner, 2003). A Canadian research group has reached similar conclusions (Furtan *et al.*, 2002).

Furthermore, the introduction of Starlink maize into the food chain without formal approval from the USDA in 2000 (Lin *et al.*, 2001/2002), the mixing of pharmaceutical maize into some 500 000 tons of soybeans by the Prodigene corporation (Gillis, 2002), and the recent accidental release of Starlink Corn by the Syngenta corporation, have raised considerable concerns within the food industry. Soon after the Prodigene affair, the Grocery Manufacturers Association and the National Food Processors Association announced to the press the concerns they had with respect to pharmaceutical crops (Simon, 2002).

Surprisingly, in a paid seminar at the annual meetings of the American Association for the Advancement of Science in 2002, Monsanto announced plans to develop pharmaceuticals in maize. (These plans appear to have been dropped later.) What was startling was the extreme lengths to which the company appeared willing to go to prevent the maize from accidentally entering the food or feed supply. This included planting in a remote area far from the corn belt, security fences and guards, video cameras, GPS, and the use of dedicated equipment to maintain the fields. The measures hardly suggested that these new technologies were harmless.

Finance

No large industry can survive in today's world without financial support. But the banking, finance, and investor communities have been at best ambivalent with respect to GM foods. As little as 5 years ago, virtually every major chemical and pharmaceutical company was investing in agricultural biotechnology research. By 2001 the situation had changed markedly. Pharmacia set Monsanto up as an independent company. Novartis did the same for Syngenta. The reasons for these decisions are fairly straightforward: when compared with pharmaceuticals, the profit potential for agricultural biotechnology was rather weak.

In addition, several years ago Deutsche Bank issued a report showing weak expectations for agricultural biotechnology (Deutsche Bank, 1999). Several years later, an investment firm, Innovest Strategic Value Advisors, reported that Monsanto was a high-risk investment (Brammer *et al.*, 2003).

Consumers

Among consumers there is still considerable discomfort about GM foods. First, consumers remain frustrated by the lack of labeling in the US. In 2002, an Oregon ballot initiative to label GM foods was defeated after the industry spent over $5.2 million. Industry officials argued that consumers would see the labels as warnings about dangerous substances. But, as *Business Week* opined, "That's shortsighted. The food industry would be better off educating the public about the safety and benefits of genetic modification. Their fear of a labeling law only means they have done a lousy job so far" (Forster, 2002, p. 44). At the very least it suggested that industry had something to hide.

Recent developments

The recent decision by the Bush administration to bring its concerns about European Union treatment of GM crops to the Dispute Settlement Process (DSP) at the World Trade Organization has once again raised the visibility of this issue to global attention (Jasanoff *et al.*, 2005). Moreover, this effort is largely futile. There is virtually no demand for GM foods in Europe as retailers there know well. Iceland Foods made headlines several years ago when it announced that there would be no GM foods in its private label products (Iceland Foods, 2003); food giant Carrefour (2002) has banned GM foods from its stores; Monoprix (2003) has replaced "products likely to contain GMOs with existing substitutable products;" Royal Ahold (2003) insists on labeling; and Heinz UK (2003) has prohibited the use of GM raw materials in its processed foods. The hostility to GM foods is similar in Japan. Thus, even if the DSP rules in favor of the US, the biotechnology industry will find it nearly impossible to penetrate the EU market.

Ironically, as the biotechnology industry has been limping along, arguably from problems of its own making, organic food sales continue to rise. Nearly every supermarket in the US and Europe has a section devoted to organic foods. Moreover, those supermarkets specializing in organic products, such as Whole Foods, are among the industry profit leaders. The same may be said for food processors. Nearly every major processor now has a line of organic foods. Even Gerber's, a wholly owned subsidiary of Novartis, promotes its organic baby foods. All this has happened without the government largesse received by the agricultural biotechnology industry. Indeed, research on organic foods remains a small fraction of overall agricultural research expenditures. But organic foods did and continue to benefit from government standards that distinguish them from conventionally produced foods and are often perceived as being healthier and more "natural."

Conclusions: lessons identified

What lessons can be identified for the new nanotechnologies from the successes and failures of agricultural biotechnologies that will help with public acceptance?

1. All actors in the innovation chain must be enrolled. Innovation is more than merely getting one actor in the product realization chain to adopt a product. The product must be viewed positively by all the actors in the innovation chain if they are not to block its use. At the very least, the new product must not result in increased costs for other actors. Moreover, those innovations most rapidly adopted are those that benefit every actor in the innovation chain.
2. Compete and cooperate. Economics textbooks tend to emphasize the benefits of competition. But good marketing and management textbooks give equal weight to competition and cooperation. Adam Smith (1994 [1776]) might have been shocked by this, given his belief that associations of all sorts would distort the marketplace. In today's world, associations, strategic alliances, contractual relationships, cross-licenses, and even co-branding, are both commonplace and necessary for success in the marketplace. In contrast, competition without regard for other supply chain actors can be and often is a dead-end.
3. Gaining by sharing. Capturing the entire market can be a recipe for failure. The tactics necessary to do that can and often do create enemies with political clout, deep pockets, a desire to invent around those who wish to control the market, and the ability to shift resources into endless legal challenges. Creating alliances is far more likely to bring success in today's networked marketplace.
4. Regulate through the entire supply chain. The US regulatory system for agricultural biotechnology is currently glued together. It is a patchwork of existing laws, none of which were designed for agricultural biotechnologies. As such the US regulatory system fails to recognize that genetically modified organisms are distributed

across the landscape by the supply chain itself through the processes of production, processing, transportation, warehousing, retailing, and consumption, as well as by the processes of nature itself. It also fails to recognize that people make mistakes, that "normal accidents" (Perrow, 1984) are bound to occur. A regulatory system for nanotechnologies should recognize this from the outset.

5. All the ducks must be lined up in order to effectively sell products. Everyone in the supply chain must both compete and cooperate. If the behavior of one actor causes damage to another, the aggrieved actor will do whatever it takes to block that behavior. Marketing of new nanotechnologies without considering this will create stumbling blocks similar to those produced in the agricultural biotechnology sector.

Endnotes

1. The current FDA regulations have the effect of making labeling of non-GM foods nearly impossible.
2. Ervin *et al.* (2000) argue that worldwide, currently available transgenic crops account for a yield increase of no more than 2%.
3. In contrast, in Canada bST is still prohibited while in Europe popular opposition continues to block use of the new technologies.

References

Brammer, M., Dixon, F., and Ambrose, B. (2003). *Monsanto and Genetic Engineering: Risks for Investors*. Innovest Strategic Value Advisors.

Brown, S. A. (1997). *Revolution at the Checkout Counter*. Harvard University Press.

Carrefour (2002). *Organismes Génétiquement Modifiés: 5 Ans d'Engagements Carrefour*. Carrefour.

Charles, D. (2001). *Lords of the Harvest: Biotech, Big Money, and the Future of Food*. Perseus Publishing Company.

Deutsche Bank (1999). *Ag Biotech: Thanks, But No Thanks?* Deutsche Bank.

Dunlop, J. T. (2001). The diffusion of UCC standards. In: Haberman, A. L., ed. *Twenty-five Years behind Bars: the Proceedings of the Twenty-fifth Anniversary of the U.P.C. at the Smithsonian Institution, September 30, 1999*. Harvard University Press, pp. 12–24.

Eichenwald, K. (2001). Redesigning nature: hard lessons learned. *New York Times* January 25, A1.

Ervin, D. E., Batie, S. S., Welsh, R. *et al*. (2000). *Transgenic Crops: An Environmental Assessment*. Henry A. Wallace Center for Agricultural and Environmental Policy at Winrock International.

Forster, J. (2002). GM foods: why fight labeling? *Business Week* **3807**, 44.

Furtan, W. H., Gray, R. S., and Holzman, J. J. (2002). The optimal time to license a biotech "lemon." Unpublished paper.

Gillis, J. (2002). Farmers grow a field of dilemma; drug-making crops' potential hindered by fear of tainted food. *The Washington Post* December 23, A01.

Haberman, A. L., ed. (2001). *Twenty-five Years behind Bars: the Proceedings of the Twenty-fifth Anniversary of the U.P.C. at the Smithsonian Institution, September 30, 1999*. Harvard University Press.

Jasanoff, S., Grove-White, R., Busch, L., Wynne, B., and Winickoff, D. (2005). Adjudicating the GM food wars: science, risk, and democracy in world trade law. *Yale J Int Law* 30, 81–123.

Kilman, S. (2001). Food industry shuns bioengineered sugar. *Wall Street Journal* April 27, B5.

Latour, B. (1987). *Science in Action: How to Follow Scientists and Engineers Through Society*. Open University Press.

Lin, W., Price, G. K., and Allen, E. (2001/2002). StarLink: where no Cry9C corn should have gone before. *Choices* 31–34.

Nation's Restaurant News (2001). Monsanto to stop producing genetically altered potatoes. *Nation's Restaurant News* April 9, 70.

Perrow, C. (1984). *Normal Accidents*. Basic Books.

Pollack, A. (2001). Farmers joining state efforts against bioengineered crops. *New York Times* March 24, A1.

Rogers, E. M. (1995). *Diffusion of Innovations*. Free Press.

Simon, S. (2002). The food industry loves engineered crops, but not when plants altered to "grow" drugs and chemicals can slip into its products. *Los Angeles Times* December 23, Part 1, p. 1.

Smith, A. (1994) [1776]. *An Inquiry into the Nature and Causes of the Wealth of Nations*. Modern Library.

Wisner, R. (2003). GMO spring wheat: its potential short-term impacts on U.S. wheat export markets and prices. Unpublished paper.

Internet references

Heinz (2003). Genetically modified ingredients. Heinz. Accessed on May 27, http://www.heinz.co.uk/

Iceland Foods (2003). Welcome to Iceland.co.uk. Accessed on May 10, http://www.iceland.co.uk

Monoprix (2003). Tableau de bord de notre démarche de progrès. Monoprix. Accessed on May 27, http://www.monoprix.fr/common/res/monoprix/fr/dev_dur_2000/tableaux/tab_1.htm

Royal Ahold (2003). Ahold Media Information. Royal Ahold. Accessed on May 27, http://www.ahold.com/mediainformation/faq/

Appendix I
A Primer on
Genetic
Engineering

We can start with a diagram of an ordinary plant, consisting of
roots, stems, leaves, and perhaps flowers (Figure A1.1). If we magnify a portion of, for example, a leaf, we are able to see the cellular
structure, the rectangular structures (Figure A1.2). In each cell is the
spherical nucleus. It is important to note that, although cells do vary

Figure A1.1   An ordinary plant, with roots, stem, leaves and flowers

What Can Nanotechnology Learn from Biotechnology?
ISBN: 978-012-373990-2

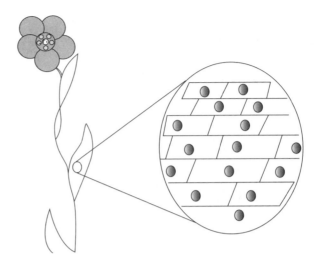

Figure A1.2   The cellular structure of the plant

in size and shape, we would see the same basic arrangement if we magnified the view of the flower, or the roots, or another plant, or even an animal, including humans. This illustrates the first concept that all organisms are made of cells and cell products.

If we magnify one of these nuclei, we see inside the nucleus an array of chromosomes (Figure A1.3). The number of chromosomes will vary from one species to another, but within a species, the chromosome number remains constant, with a few exceptions. Humans have 46 chromosomes; a wheat plant has 42. The number of chromosomes in any given cell of an organism is the same, whether we count them in leaf cell, flower cell, or root cell. So while the number may vary from one species to another, the number in each cell of a given individual is constant, and common to that species.

Another step of magnification, looking inside the chromosome (Figure A1.4), shows the DNA, the molecule responsible for storing hereditary information in all living things. DNA is the ubiquitous long thread molecule packed into the chromosomes of every cell of plants, animals, and microbes. The DNA of a single human cell, if extracted and pulled taut, would reach about 2 m in length. The DNA from a single cell of a small plant might be only a few centimetres. The total amount of DNA is not related to a species complexity; some species (e.g. wheat and frog) have DNA even longer than that of a human cell.

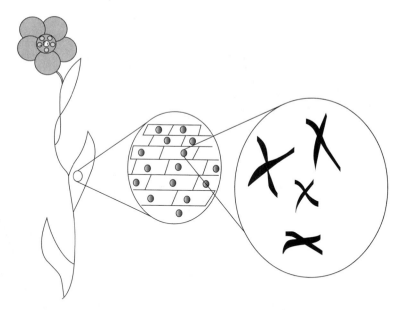

Figure A1.3   Chromosomes

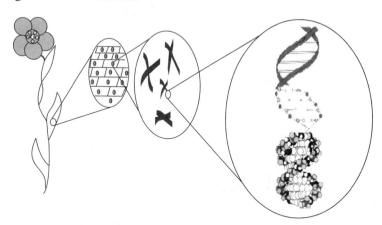

Figure A1.4   DNA

How is hereditary information stored in DNA? To answer this, we move to a more schematic magnification to investigate DNA.

If we consider a short piece of DNA and pull it taut, we can conceptualize how genes, the units of hereditary information, are arranged along the DNA molecule.

Here, we simply designate stretches of DNA as gene 1, 2, 3, 4, or 5 (Figure A1.5), with some intervening portions of DNA not associated

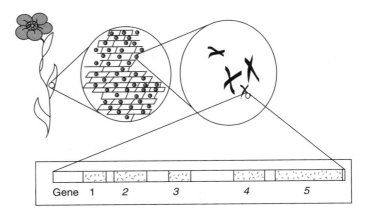

Figure A1.5 Genes

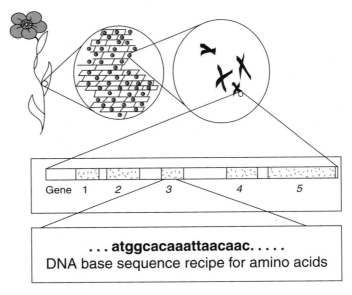

Figure A1.6 Sequence of chemical bases making up the DNA within the gene

with any gene. The genes are the hereditary recipes, passed down from generation to generation, just like recipes are in some families. How do these stretches of DNA store information? Let's look more closely at a portion of DNA for one of the genes.

The magnified view of DNA at the left edge of gene 3 (Figure A1.6) shows the familiar ATG and C chemical bases (actually abbreviations for the chemical names), the chemical building blocks which

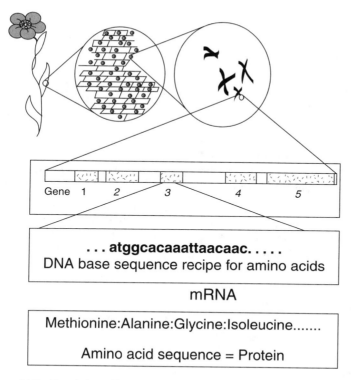

Figure A1.7 Translation of base sequences into amino acids

make up all DNA molecules. The DNA can be extended by adding more of these bases, or reduced by lopping some off, hence the variation in length of DNA in cells of different species. The particular sequence of bases conveys the information to the machinery of the cell, which then builds a particular protein, conceptually the same way a food recipe provides directions to compile a particular meal. Continuing the recipe analogy, the four base "letters" in DNA language are analogous and comparable to the 26 "letters" in English. The letters in each language can be arranged in a particular sequence to give meaning. In English, the specific sequence of letters builds words and sentences; in DNA the letter sequence spells out a call for particular amino acids and the construction of a protein composed of a specific sequence of amino acids.

For example, Figure A1.7 shows the DNA base sequence at the start of gene 3 is ATGGCA. … In DNA language, all "words" are three letters long. The first word in this sequence, ATG, translates as the amino acid methionine. The next word is GCA, which calls for the amino acid alanine. When the cell machinery reads gene 3, it follows the recipe by finding a methionine floating around the cell, and connects to it an alanine molecule. The next word, CAA, calls for glycine, so the cell finds and attaches a glycine molecule to the methionine and alanine already connected, thus building an elongating chain of designated amino acids. An average gene is about 1000 bases long, or 330 genetic "words," which means a protein chain consisting of about 330 amino acids. The particular sequence of amino acids gives the protein its particular features, which in turn give the cell and ultimately, the organism, a particular trait. Some proteins are very short, simple chains of amino acids, others are long, complex, composed of several amino acid chains or are further modified prior to activation. Human insulin, for example, is a fairly simple protein composed of two amino acid chains, one is 30 amino acids long, the other 21 amino acids long. (Many other animals also generate insulin. Some, like rat insulin, is almost identical to human insulin. Until recently, human diabetics were treated with bovine or porcine insulin, as it is similar enough to human insulin to do the job.) In any given cell, there are approximately 10 000 different proteins. Many are enzymes, helping perform various chemical reactions.

In genetic engineering, the DNA carrying the gene recipe for a desired trait is copied from an organism that has the gene and is conveyed into the plants (or other) cells on a small circular piece of DNA called transfer DNA, or t-DNA. There are several methods to deliver the t-DNA to plant cells, in each case the natural cell mechanism is responsible for inserting the t-DNA into the native DNA. The exact technical mechanism is unclear. However, the process is a cellular one, in that the t-DNA is delivered to many cells at a time, but only some cells successfully integrate the t-DNA into their own genome. This is illustrated in Figure A1.8, showing the t-DNA being inserted into only one cell in the plant leaf.

If we now magnify that "transformed" cell, we can conceptualize the inserted t-DNA (abbreviated i, for insert) as having been integrated between gene 2 and gene 3 (Figure A1.9).

Now, when we magnify and read the inserted DNA (gene i), we see the same four DNA bases (ATC and G), but the sequence is different from that of gene 3 (Figure A1.10).

In this case, gene i starts with the same ATG as gene 3, but then has bases GCC and subsequent bases are also different (Figure A1.11).

When gene i is read by the cell machinery to make a protein, the amino acid sequence chain will be methionine, alanine, leucine,

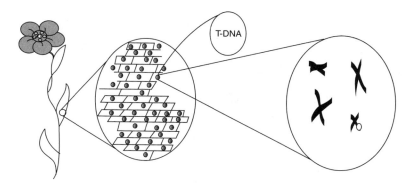

Figure A1.8 t-DNA inserted into one cell in the plant leaf

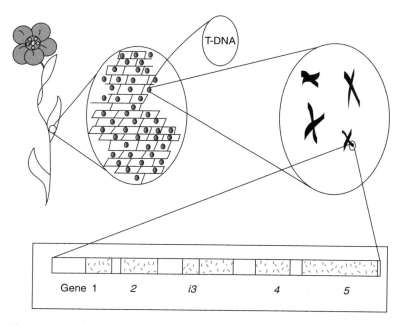

Figure A1.9 The t-DNA (gene i) is integrated between gene 2 and gene 3

tryptophan and so on. This is a new amino acid sequence, resulting in a new protein, providing the plant with a new trait, depending on the function of the new protein. The new trait may confer resistance to a disease, or it may provide a new nutrient, or a new industrial protein. The gene itself, the physical piece of DNA, is simply baggage in the transfer procedure.

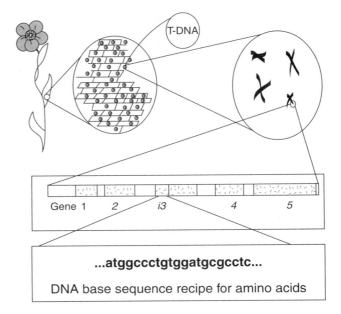

Figure A1.10   Base sequence in gene i

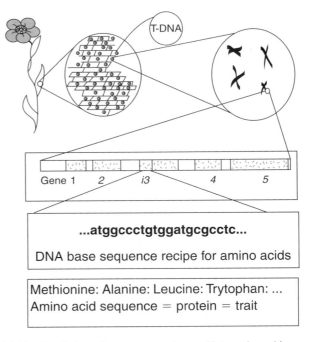

Figure A1.11   Translation of base sequences in gene i into amino acids

Appendix II
Report from the Standards for Nanotechnology Workshop

Objectives: highlighting issues for nanotechnology standards

Research, development, and commercialization of nanotechnologies are moving forward in a period of uncertainty about standards and regulation. This uncertainty poses different opportunities and constraints for a variety of stakeholder interests across sectors of economic activity.

The Agrifood Nanotechnology Project at Michigan State University (ANP-MSU) is supported by a National Science Foundation (NSF) grant to examine social and ethical dimensions of nanotechnologies in the agrifood supply chain. The Agrifood Nanotechnology Project is jointly conducted through the Department of Community, Agriculture, Recreation and Resource Studies (CARRS) and the Institute for Food and Agricultural Standards (IFAS) at Michigan State University (MSU).[1]

ANP-MSU activities include research, educational development, and convening international conferences. From issues that became apparent at the What Can Nano Learn from Bio? Conference that convened on October 26–27, 2005 at MSU, the Agrifood Nanotechnology research group designed a follow-up conference.

What Can Nanotechnology Learn from Biotechnology?
ISBN: 978-012-373990-2

The subsequent workshop, Standards for Nanotechnology, convened September 11–12, 2006 at MSU. Both the workshop and this report aim to identify—from a variety of stakeholder perspectives—key issues for standards development (IFAS, 2007).

Defining terms and key issues

It should be noted that in the US, the term "standards" is applied to both voluntary standards set by various private and/or non-profit organizations as well as mandatory regulations set by government agencies. On the other hand, in the EU voluntary standards are usually contrasted with government regulations. In recent years, due partly to increased international trade, the distinction between standards and regulations has become blurred. Many voluntary standards have become de facto mandatory standards. In this document we follow the broader US usage of standards and distinguish between voluntary and mandatory requirements when necessary.

Basic research is needed to determine the health, safety, and environmental impact of emerging nanotechnologies. Without such data, it is difficult to move the standards-setting process forward. Additionally, common nomenclature and cooperative frameworks need to be established early in the process of technology development. These are overarching issues. The remainder of our discussion and recommendations clusters around five key issues. We briefly introduce each issue.

1. Timing and standards-setting

Standards for nanotechnology need to be developed for all of the stages in the life cycle of the products (research, production, products, waste, etc.). Research into nanotechnology is already moving forward under existing standards for lab safety, but development of nanotechnology-specific standards is needed. The production phase is also likely to be a high-risk point. Agencies experienced with worker health and safety should be engaged early. Standards to regulate consumer products are also lacking, and there is disagreement about whether new legislative authority is needed to guide the elaboration and implementation of such standards. Finally, standards for nanotechnology lab waste, production waste, and end-of-product-life waste raise new questions. Both private and governmental actors should collaborate to address these issues. We urge prioritization of these areas based on the most current safety and risk data, as well as adequate funding for risk analysis to ensure that standards-setting is able to keep pace with research and development.

2. Product vs. process standards

Production processes at the nano level are not new. Catalytic conversion process using nano-scale supports for catalysts have been employed for at least 30 years (see Chapter 10). By contrast, nano-products are largely still in development. Are both process and product to be regulated in the same way? Nanotechnology raises questions about where within the life cycle of a product it makes most sense to place various standards and regulations. Nanotechnology risk assessments and analysis will be useful in determining the most efficacious ways to implement these standards and regulations. Different government agencies have different mandates and this will likely have a large impact on whether we have standards set for products or processes. We encourage interagency cooperation to create the most effective standards. Agreement is needed on the goals of the setting of standards to clearly decide whether it makes more sense to regulate products or processes.

3. International harmonization

The US, EU, and Japan are all heavily investing in nanotechnology development. Given the global economy, it is certain that intermediate nanotechnology products and finished goods will be marketed globally. This calls for limited international harmonization of standards and regulations. Dialogue and cooperation among diverse stakeholders is needed to determine which standards should be harmonized and how enforcement of international standards should be carried out. We recognize that the debate over international harmonization of standards for any technology is a debate about the different concerns and priorities a nation gives to worker health and safety, environmental protection, economic competitiveness, etc. A certain level of national autonomy in these realms is reasonable.

4. Integration of operational standards

The development of effective nanotechnology standards will require that standards-setting agencies that have not historically worked together begin to do so. Mechanisms for interagency cooperation should be primary goals in this process. Achieving information sharing and effective interagency communication will serve as first steps towards more effective standards-setting. Lawmakers can assist in this process by providing adequate funding and clear authority to integrate agency mandates. Careful consideration is needed in choosing an appropriate model. Top-down models should be avoided. Instead we suggest that bottom-up models should be explored. The Coordinated Framework for Biotechnology was established in 1986 by the US Government to regulate biotechnology products by recommending mechanisms for interagency coordination. This framework provides a possible, though limited, initial model for nanotechnology. Where appropriate, integration with the private sector is also recommended, as some private sector bodies now act as de facto standards-setting bodies. ISO, Codex, and the IPPC offer good models for this kind of integrated approach.[2]

5. Participation and transparency in standards-setting processes

It was once seen as acceptable for scientific experts, government bureaucrats, and businesses to debate and establish standards with little or no input from the public. This approach has justifiably been called into question in recent years. In the future, we expect to see more attempts at public participation in the standards-setting process than has previously been the case. The type and nature of public participation is largely undefined, and it is this area that needs the most attention. For the public to consider its participation as legitimate, careful attention needs to be paid to identify potentially affected groups and engaging them in meaningful ways in the standards-setting process. Several models for this were discussed. Standards-setting bodies need to review and learn from models that have been more successful than the typical "public meetings" model common in the US.

To summarize this opening statement, the purpose of the standards workshop and this report is not to establish consensus around these themes, but rather to chart the "issues landscape" facing the nanotechnology standards communities. As such, this report serves as a roadmap to inform the standards deliberations of agencies and organizations confronting emerging nanotechnologies and their potential applications both within and across different sectors of economic activity.

Background: challenges for nanotechnology standards

Various new nanotechnologies have been singled out by their proponents as unique, but this uniqueness poses problems for standards. Even the development of the nomenclature for the description of the new nanotechnologies is a complex task. There are currently several organizations involved in the development (e.g. International Organization for Standardization [ISO] and American National Standards Institute [ANSI]). Historically, much standard-setting has been reactive in response to injuries, while current efforts involve a more proactive approach. On the one hand, we may need reactive standards (i.e. standards for reporting negative incidents), similar to those currently used in the food industry to report food safety problems. On the other hand, anticipatory standards are more desirable. However, these would almost undoubtedly have to be linked to particular products. It is extremely difficult to develop an effective standard proactively without a specific product in mind.

In the US, attempts to coordinate federal work on the nanoscale began in November 1996, when staff members from several agencies held formal meetings under the auspices of the National Science and Technology Council (Stone and Wolfe, 2006). In 2001, the Clinton administration raised nano-scale science and technology to the level of a federal initiative, officially referring to it as the National Nanotechnology Initiative (NNI). The NNI now coordinates the multi-agency efforts in nano-scale science, engineering, and technology under the "21st Century Nanotechnology Research and Development Act" (108 P.L. 153, 2003). Twenty-three federal agencies presently participate in the NNI, 11 of which have research and development (R&D) budgets for nanotechnology. The "Supplement to the President's 2006 Budget" (NNCO, 2005) recommends overall NNI investments for 2005/2006 of about $1.05 billion, with $82 million devoted to "societal dimensions" including "environmental, health, and safety R&D" ($38.5 million) and "education and ethical, legal, and other societal issues" ($42.6 million). The NSF, one of the agencies participating in the NNI, sponsors a number of nano-related priority areas, such as nano-scale exploratory research, nano-scale interdisciplinary research teams, and nano-scale science and engineering centers. In the fiscal year 2005, total funding for these NSF programs exceeded $296 million. Recent government projections suggest that funding for nanotechnology will continue to rise across all sectors, with global expenditures projected to exceed $1 trillion by 2015 (Roco, 2003). The NSF is currently the major source of federal funding for research related to nanoscience and nanotechnologies. The Agrifood Nanotechnology Project at MSU, through which the International Nanotechnology Standards Workshop was convened, is supported by an NSF grant to examine social and ethical dimensions of nanotechnologies in the agrifood supply chain.

Addressing the challenge

Standards are generally considered to be convenient, neutral, and benign means for handling issues of technical compatibility. However, if one thinks of social power as the ability to set the rules that others must follow, then standards represent a form of codified power reflecting the interests of those groups with greatest access to and influence within standards-setting processes. While many people and institutions recognize and broadly support the role of standards in general, controversy often ensues as they confront the question: "Whose standards?" We recognize that standards are shaped by cultural, ethical, political, and strategic, as well as technical considerations, and this perspective guides the standards-related activities associated with the Agrifood Nanotechnology Project.

The 2006 Standards for Nanotechnology Workshop

The ANP-MSU project is engaged in research and outreach activities pertaining to the development of nanotechnology standards. For example, our project team presently holds a seat on the ANSI Nanotechnology Standards Panel, from which we are able to observe and to some extent participate in the standards facilitation process. Some of these activities relate to standards for other technologies as well as with standards regimes operating across specific spheres of economic activity—such as global agrifood supply chains. The ANP-MSU has also sponsored a series of nanotechnology conferences and workshops. The first of these, convened in October 2005, examined experience with agrifood biotechnologies as seen from a variety of stakeholders—integrating the lessons learned from this prior experience to inform the ongoing development of nanotechnologies.

Of the many lessons learned, the most prominent was the perceived failure to engage diverse stakeholders and other potentially affected groups in a dialogue as standards for agricultural biotechnologies were being set. Conference participants agreed that an early dialogue among diverse stakeholders should precede the development of standards for emerging nanotechnologies, so as to better identify the social landscape and potential consequences of standards decisions. The 2006 workshop was convened to bridge the interests of a variety of communities that might not normally communicate with each other on standards issues, but which nonetheless maintain a mutual interest in them. The goals of the workshop reported here were to:

- stimulate public discussion and understanding of issues involved in developing nanotechnology standards,
- influence public and private agendas with respect to nanotechnologies, and
- link diverse and distinct communities concerned with nanotechnologies.

Although the workshop addressed standards issues likely to be relevant for the agrifood sector, emerging nanotechnologies are expected to cut across numerous sectors, potentially blurring traditional boundaries such as food, pharmaceuticals, and cosmetics. ANP-MSU invited workshop participants representing diverse perspectives, including business and industry, government regulatory agencies, labor groups, non-governmental organizations (NGOs), trade associations, and standards-setting bodies, as well as numerous domestic and international academic and technical disciplines. Prior to the workshop, the ANP-MSU created a web forum for participants and interested invitees to begin online discussion on each of five critical standards themes. They were able to shape the workshop agenda more carefully and ensure that it addressed their concerns. The five key issues themes were:

1. Timing and standards-setting
2. Product vs. process standards
3. International harmonization
4. Integration of operational standards
5. Participation and transparency in standards-setting processes.

The workshop was designed to maximize participation and enlist diverse perspectives.

At the workshop, participants were divided into five small working groups to facilitate discussion. To maximize variation of perspectives, each group consisted of members from each of the stakeholder categories identified above. Each session of the workshop started with breakout groups identifying and discussing questions and issues surrounding one of the five critical standards issues. Project team members facilitated these breakout groups; there was a note-taker for each of the groups. Each breakout session lasted roughly 2 hours. Then the groups reconvened to report the results of their deliberations.

During the following months, the project staff compiled notes from each breakout group and organized them around each of the five standards themes. These notes were drafted into text reflecting key issues and questions within each theme. Early drafts of this report were posted on the workshop web forum for participant review and comment. This process helped to further clarify key issues and questions. This document synthesizes ideas focused around each of the five themes. The purpose of this exercise was not to establish consensus around these themes, but rather to chart the "issues landscape" facing the nanotechnology standards communities. As such, it is designed to serve as a roadmap to inform the standards deliberations of agencies and organizations confronting emerging nanotechnologies and their potential applications both within and across different sectors of economic activity.

Workshop results

The remainder of this report is a summary of points raised by the workshop participants when discussing the five key issues.

1. Timing and standards-setting

With the rapidly emerging field of nanotechnology, it is important to balance the need for free inquiry with the need to protect society through the development of standards. Standards are needed for basic research laboratories, product development laboratories, and manufacturing facilities. Environmental standards and consumer health and safety standards are needed.

When should discussion and identification of standards for new technologies begin? This is a critical point of debate between scientists, governments, NGOs, industry, and the general public. Discussion centered on three interrelated themes: (1) beginning the standards-setting process for nanotechnologies early in the knowledge development process, or later as such knowledge is applied to the development of new products and processes; (2) developing a timeline that is acceptable to all interested parties; and (3) addressing these issues appropriately and strategically with respect to global economic competition. The following is a synthesis of ideas surrounding these themes. Standards will need to be developed for such aspects of the new nanotechnologies as research, production, products, and waste disposal. At the same time, it is likely that standards will vary by material and product.

Research

Research standards should initially entail the use of Good Laboratory Practice (e.g. special gloves, respirators, and hoods). Many gloves used today in conventional lab work are ineffective for preventing penetration by nanoparticles. Some companies and universities already employ standards for Good Laboratory Practice. However, fully eliminating exposure might have the effect of creating a de facto ban on nanomaterials. Since university-based research is often conducted further upstream from industrial research, it may require standards that are different from those for industrial research. At the same time, virtually all universities are faced with declining resources for facilities. We need to ensure that researchers, post-docs, and students are adequately protected from what remain largely undefined or poorly defined hazards. Given the scarcity of information about risks associated with nanotechnologies, initial information and insight should be collected based on experiences in the laboratory. This suggests that there is already a need for standards for reporting incidents of concern, for providing guidance on what kinds of incidents should be recorded as "potential negatives," for reporting certain incidents should be reported to pertinent agencies.

Production

Standards are needed to protect workers from exposure. The highest exposure risks are likely to occur during the production phase, not in the final products. There are means for reducing risk during the production phase. If risk is a function of exposure and hazard, and we do not know what the hazard is, then we should at least reduce the exposure. The National Institute for Occupational and Health has a good track record in working with industry in similar situations. Its expertise should be welcomed.

Products

Indeed, one problem posed by the new nanotechnologies is that every new product seems like a special case. Moreover, different products will fall under the jurisdiction of different regulatory agencies. For example, a product which is overtly therapeutic will fall under the FDA approval process and require clinical trials. Food and drug products will likely need standards as soon as products are developed. In contrast, there is less urgency for non-consumable products, although environmental issues still need to be addressed. Even the determination as to whether a new food or drug product is substantially different requires a standard. Furthermore, there are naturally occurring nanoparticles in our food and in other consumer goods now. We will need to differentiate between these naturally occurring nanoparticles and those that are manufactured. One issue of considerable importance with respect to products will be public disclosure. Many, if not most, nano-products will involve Confidential Business Information (CBI); therefore, only the commercial applicant's lawyers' summaries (of risk relevant data, for example) will be available to the public. Currently, no pre-market notifications are required for nano-products. According to the manufacturers, nano-tubes are in tennis rackets now. Labeling may pose yet another set of problems. The insurance industry is likely to have a significant effect on the use of nanotechnologies in consumer products. Currently, insurers are uncertain about their capacity to insure against damages resulting from the production or consumption of products with nanomaterials due to the lack of standards, and their inability to calculate their actuarial exposure to economic risk. One effect of this uncertainty will be pressure on firms using nanotechnologies to self-regulate in order to avoid tort cases.

Waste disposal

Once nanoparticles are brought together in a product such as a tennis racket, do they come out again? These particles pose a unique end-of-product-life concern. Virtually all industries have environmental discharges, and many engage in wastewater treatment. Currently, nanoparticles are not differentiated in industrial processes, so waste streams are not separated. Moreover, the liquids in which nanoparticles are stored may be demonstrably more toxic than the particles themselves. Furthermore, because filtering processes were intended to catch much larger particles, it is likely that current waste treatment procedures and processes are not sufficient to filter out nanoparticles. Given the differences in the behavior of chemicals at the nano level, it is conceivable that the Environmental Protection Agency (EPA) will need to re-review every chemical in its database. The task will be daunting, and one for which funds are currently lacking. This suggests that standards need to be developed now to prioritize according to toxicity and likelihood of use.

2. Product versus process standards

In many areas of new technology development, debates have arisen as to whether standards should focus on process or on product. For example, standards for organic foods are process standards prescribing particular production processes to be followed in order to meet standards, while those for pesticide residues are product standards defined by the quantity of pesticides remaining in/on the food product at the point of consumption. Similarly, one can distinguish between standards for products bearing nano-engineered materials (e.g. new quality attributes, safety issues, etc.) and those for production processes and management systems (how nano-devices are made, and also for nano-enabled processes used to generate new or modify existing products bearing no nano-engineered materials).

Discussion centered on three interrelated themes: (1) whether standards for nanotechnology should be primarily product-based or process-based; (2) the extent to which the answer to this question is dependent upon (or likely to vary according to) the specific nanotechnology in question; and (3) whether such standards should vary by the intended application both within and across particular sectors of economic activity (e.g. agrifood, medicine, energy, security, etc.).

Definition of product and process

It was felt that the general answer to this question depended on the definition of process. Process could refer to an engineering process, production process, or a governance process. Nanotechnology is such a broad category that pinning down "nanotechnology product" or "nanotechnology process" might be like trying to set a standard for products/processes of biology.

It is similarly difficult to define the product. Is it what the consumer touches, or is it the primary materials, before they are purchased and used by a consumer? It was suggested that a standard might need to be developed to determine what is a nanotechnology process or product, and that this would help avoid too imprecise a definition of nanotechnology. This might require the creation of new language and new nomenclature. There could also be triggers established for when nano-standards would be used. Currently, some products might be missed because they are not labeled "nano."

Some participants questioned whether processes at the nanoscale are sufficiently unique as to require process standards. Natural scientists noted that some processes in nanotechnology are well established and the products are quite ordinary (see case of Zeolite Catalysis in Chapter 10). Other processes are well established although their products are novel. It might be much easier to have standards for products, given that there are so many ways to process things. But some processes might be seen as unacceptable, even though the product produced by that process is acceptable. Workers can be subjected to considerable risk, even as the products they produce meet all health and safety standards. Moreover, these concerns are not limited to health and safety.

Discussion of examples seems to blur the process/product line. Organic standardization was the touchstone for process standards. Meat standards blur the line, since often the kind of product (e.g. veal) is directly linked to the process. Milk, meat, and egg standards show that both process and product are currently standardized.

Risk

The assessment of various nanotechnologies may force a prioritization of risk with respect to nanotechnology standards. Part of this assessment will reveal if, or when, there are special impacts of nanotechnologies. Decisions to address product or process might emerge from the risk assessment of a given nanotechnology. At the limit, addressing issues of risk will require thinking about standards in a manner that is much broader than risk assessment itself.

Agency interaction

There was much discussion regarding which agencies might focus more on product standards and which on process standards. The Joint FAO/WHO Expert Committee on Food Additives might be asked to assess risks of nanoparticles in foods, or in food packaging. Regarding the certification of organic foods, the FDA and the Organic Crop Improvement Association might be likely to regulate by process, whereas the EPA might regulate products. These agencies might best start with an existing process or product standard, and then extend that to related nano-products or processes.

This viewpoint was questioned: it was pointed out that existing standards may appear inadequate to include new techniques and products. The Occupational Safety and Health Administration is approaching nanotechnologies with some consternation. We must avoid the myth that is widespread in the biotechnology debate: that the US only regulates products, and the EU only regulates processes. Decisions regarding whether to employ product or process standards may vary by sector of economic activity, reflecting differences in legislative mandates across sectors.

The process/product question raised the issue of how to audit/monitor process standards. Various strategies were suggested: third party auditing, accreditation, and certification. Third party is favored for control, but raises concerns about cost and limits on producers' freedom to operate. Post-market monitoring remains a huge issue.

Goal-driven standards

Part of the difficulty in deciding between process and product standards has to do with the goals of standardization and those of particular nanotechnologies. Are we trying to protect consumers, protect workers, or limit certain kinds of commerce? Is the goal to reassure the public? The public seems more reassured by product-specific standards rather than those related to processes. It was thought that a product-only standard environment would result in de facto privatization of process standards. It is clear that some smaller and start-up organizations struggle due to their inability to get access to standards-setting processes.

An important divide emerges. On one side is the view that process standards are concerned with values, and that product standards relate (more objectively) to safety or health. The opposite view emerged that process and product themselves are so linked that one cannot separate value concerns from health/safety concerns. Furthermore, the cost of standardization will enter into decisions about when, where, and if product or process standards are used. Many supported the view that process standards are harder to monitor and would therefore be more costly, especially in a third party auditing situation. Standards will probably vary according to the specific nanotechnology in question.

Ordering and goals of standards
Standards for nanotechnologies may be approached from three directions: (1) Specifications for the production of the nano-device, and the engineering practice of producing the nano-product; (2) How that nano-product is integrated into other production supply chains; and (3) How nanotechnologies are integrated into the product itself.

Convergence and jurisdiction
One issue that is relevant to this question is the blurring of lines between sectors because of nanotechnology's convergent nature. The processes and products of nanotechnology cross traditional sector lines. One way to deal with this is to stipulate that any relation to nanotechnology requires the process or product to be listed as a nanotechnology. Another variant that may determine the extent to which product/process standards are used is jurisdiction. States and countries will employ different approaches, a strategy generally seen as desirable. For example, Brazilian experience with biotechnology has shown that a lack of negative connotations for nanotechnologies may push some nations to regulate only products, instead of processes (Mattoso et al., 2005; Rattner, 2005).

Transformation of society
Some humanists and social scientists argue that the social dynamics of nanotechnologies are completely different from earlier technologies and they will bring about profound social changes. Doubtless, process/product standards will be part of a transformation of society with respect to goals, regulation and technology. There are people

talking about a "new industrial revolution." This was seen by some participants as an unacceptable goal if it merely means that a few people will profit in an unregulated environment. The philosophical issue of enhancement vs. medical application arose with respect to this sector issue. It is unclear whether some procedures are enhancing or treating. Are technologies solving problems or "just" improving things in a more cosmetic way? This may affect the decision to regulate process over product.

3. International harmonization

The US, EU, and Japan, among other nations, have invested significantly in nanotechnology development. Given the large and growing global trade in raw materials, intermediate, and finished goods, it is more than likely that products produced using nanotechnologies and products incorporating nanotechnologies will enter into international trade. Furthermore, it is likely that some nanotechnologies will require harmonization and/or interoperability of standards across national boundaries. While several countries are very active in nanotechnology standards development, little discussion has taken place regarding their global harmonization. Discussion centered on four interrelated themes, including: (1) the kinds of standards that will need to be harmonized globally; (2) how to ensure that the interests of countries other than those identified above are included in global standards harmonization; (3) whether certain standards can remain local/national in scope; and (4) preferred ways of moving the process of international harmonization forward.

From national to international
Before discussing what kinds of voluntary standards and government regulations will need to be harmonized globally, we should deal with the concern about standards proliferation, especially when many countries lack adequate enforcement capacity. It is dangerous to do so. At the same time, inadequate enforcement should not limit the drive toward harmonization, but should be coterminous with building capacity. For international harmonization to be possible, it is necessary to first deal with national standardization. Each country still has obligations to its own population. Harmonization will be difficult because it must be compatible with many diverse cultures. How to deal with these cultural differences while constructing standards is

unclear. The identification of priority areas will be difficult since public concerns will differ from country to country. Participation in the forum that ISO provides is critical. We fail to participate at our peril. In the agriculture and food sector, Codex and/or the International Plant Protection Convention are more relevant. ISO has experience in developing product, nomenclature, process, and test standards. Technical standards will be needed to determine whether particles are found in a product and can be released in a consequential manner. Within the US, participation in ANSI deliberations should be encouraged. Some participants noted that the ISO process that allows one vote per country can be very political. Thus, it is not always possible to develop harmonized standards using this format. Alternative approaches to international standards-setting should be considered. For example, the American Society of Mechanical Engineers (ASME) sometimes functions as a default international standards-setting body, but is not always recognized as such. Harmonization will also require communication among different bodies. This will raise the issue of data privacy and transparency, especially since some nanotechnologies (e.g. radio frequency identification devices [RFID]) will enhance the ability to transfer data. Harmonization of information gathering for processes and products will be needed to establish consistent methods. Furthermore, if there is too much pressure to keep the cost of standards low, then there is insufficient cost recovery to allow for the constant development of new and evolving standards. This issue needs to be addressed. Standards are often copyrighted or otherwise protected by standards owners (e.g. ASME standards). This can be a barrier to adoption. There are cost barriers associated with ISO, ASME, ASTM, and other standards.

Which standards?
In harmonizing standards at a global level, there was considerable agreement that the main focus needs to be on public health impacts. Labeling, quality issues, and environmental issues also need to be harmonized internationally. Worker safety is a more complex issue as standards-setting works differently in different places and organizations. However, standards need to be harmonized where they concern risk, exposure, and waste disposal. Product standards do not necessarily need to be harmonized but perhaps process standards do. If nanotechnology standards evolve from current standards, then there will be a combination of national and international standards.

It might be possible to begin by agreeing on principles for standards rather than on specifics. International standards have more potential to become politicized while national standards can be developed in a manner that is relevant to local conditions. Worker health and safety standards are complex. For example, some participants observed that high worker standards are an economic disincentive. This might be a reason to keep this local or national in scope. In nanotechnology applications, if there are environmental consequences, they must be related to local and national situations. However, nanotechnologies exhibit unique features and do not have national boundaries. Some nano-products could have international implications if they are released into the atmosphere. Therefore, the question of the right of a country to refuse to be in contact with the product needs to be addressed. There also is the issue of the right of a government to refuse exposure of its citizens to certain materials.

Which countries?

In multilateral standards-settings processes, the exclusion of some countries for practical reasons is a limit to international harmonization. Also, the cost of the standards can make them unavailable to developing countries and prevent them from participating at the international level. Consequently, cost recovery is an important issue. Developing countries should have a say in international nanotechnology standards development, even if they lack capacity to enforce the standards. This is important because standards often control whose exporters can enter a given market. One means to address this might be for public agencies in industrial nations to set aside research or intellectual property for developing countries as a "tax." Regional discussion might also help to strengthen the position of developing countries at the international level. The most important thing is to establish local and regional standards and then to navigate international barriers. Finally, given current concerns over national security, some participants asked whether inclusion of all nations was desirable.

4. Integration of operational standards

Nanotechnologies pose new challenges for operational standards. They must ensure that the health and safety of workers and consumers are protected and that environmental protections are developed and enforced as needed. In the past, these issues were the subject of separate regulations and regulatory agencies (e.g. in the US, OSHA, EPA, FDA, and USDA), with each holding responsibility for different aspects of the regulation of agriculture and food products. Integration of diverse standards regarding nanotechnologies is likely to pose new challenges for governmental regulation and non-governmental standards. Discussion centered on three interrelated themes, including whether: (1) both private and governmental standards-setting agencies that have not historically worked in cooperation will begin to do so, (2) procedures can/should be established to ensure adequate integration among different standards agencies, and (3) unique challenges will be raised by this issue, and if so, how to identify the best strategies for addressing them.

Integration among agencies

In some circles there is an assumption that agencies will just start to work together as the need arises, but this is not likely to happen on its own. Instead, mechanisms must be established for interagency cooperation where it has not occurred before. Success in this endeavor is key, so we need to look for examples where different agencies have worked together cooperatively on standards and regulation in the past. Currently, in the US there is very little or no interagency talk on emerging nanotechnology standards and regulatory needs. For example, the EPA recognizes that there will be nano-waste to regulate, but they do not know much more than that. They do not know what the waste will be or how much of it there will be. There needs to be integration with other agencies that can help them understand this, and the integration needs to go beyond other regulatory agencies. Another good example is nanosensors and RFID tags, where there may be a need for the USDA to integrate their standards with those of the Federal Communications Commission. This probably has never happened before. Still, other agencies probably are not thinking about nanotechnology at all. For example, we doubt that Grain Inspection, Packers, and Stockyards Administration have considered the importance of nanotechnology regulation. Likely, they will need to do so. Therefore, mechanisms will need to be established to help them begin to see their role in this regulation. Compounding this, is a history of non-communication among regulatory offices. We need to foster better communication methods to overcome this history. A good place to start might be with integrated computer systems. A big challenge is information sharing, so the easier and more efficient information sharing between the agencies can be made, the better. Designing a central database appears to be a useful first step. There is, however, some tension between agencies when it comes to information sharing (e.g. who gets the credit?, etc.). There needs to be a focus on overcoming this. Oftentimes, barriers occur at a specific point. If communication can be facilitated at those points, the entire interagency process can be enhanced. Furthermore, each individual agency may have insufficient funds to accomplish this on its own. Additional funding will be extremely important to success. Ultimately, two things are needed: someone with the authority to integrate and an agency to catalyze the process. In the US, some entity other than the NNI should play this role, although its NNCO does have a similar function within the NNI. However, creating a completely new agency could conceivably generate more confusion. The bigger question remains which agency would take the lead and act under what mandate?

In this discussion, there was a conflict between top-down and bottom-up approaches to integration. However, it is difficult to provide an example of a top-down approach that has worked well. The Department of Homeland Security has a top-down model with enormous power and vast resources, yet its success is questionable. Therefore, bottom-up models of integration are needed. Moreover, the Coordinated Framework for Biotechnology (OSTP, 1986) should be examined to determine if a similar approach would work with a larger number of agencies.

Integration with private sector

Five of the top 10 food companies are major nanotechnology investors. These and other private/corporate investors have a great deal of information that is likely to be useful to the standards-setting process. Integration among agencies and also among all supply chain actors is also important. Wal-Mart is now a de facto standards-setting body for quality standards. Standards and regulations have historically ignored the complexity of the supply chain. Yet, integration with private sector players could help establish new approaches similar to hazard analysis and critical control points, commonly used in the food industry, along the supply chain. Instead of waiting until a product is complete, one can test it along the way. At the same time, we should be wary of the role that people/companies with a financial stake in nanotechnology play. Various US government agencies are collaborating with ANSI and ISO in standards development. ASTM's E-56 committee is also involved in nanotechnology standards development, and has recently released its standard "E-2456, Terminology for Nanotechnology" (ASTM Committee E-56, 2006). There is considerable overlap in membership among these and other standards organizations, and they may serve as a fruitful place to begin the discussion of standards integration. Finally, global integration will require cooperation among competing institutions. Typically, the tension that results from competition limits cooperation on regulation. Additionally, who integrates with whom becomes a point of contention.

Challenges: old and new

The greatest challenges are not so much unique, as they are persistent and unsolved. Such is often the case with new technologies. If we can solve the problems presented here, we will have taken a large step towards solving the regulatory challenges associated with all emerging technologies. These challenges lie largely in the complex social dimensions of technology. The proposed national animal identification system is a good example. Regulatory agencies thought this would be a simple matter of organizing a database and implanting tracking devices. However, it quickly became a larger social issue involving government knowledge of herd size, location, transport, etc. Also, with such a database a disease outbreak beyond a farmer's control can be traced back to the farmer. This poses potential issues of liability and social stigma. Yet another challenge to nanotechnology is that there are already products on the market that need to be reviewed for safety and efficacy by regulators, but it is not obvious who should do this. With nanotechnologies it will be important to set standards in parallel with product development so as to avoid this in the future. Also, nano-waste that has not been evaluated for health, safety and environmental risks already exists. Companies are unlikely to use their R&D funds for this type of research; a governmental body needs to do it, and sooner rather than later. In addition, the hazards associated with nanotechnology are largely unknown. This makes it difficult to assess how to proceed. Standards are needed for working in situations of uncertainty. Some form of the precautionary principle might be appropriate, at least until the hazards are more well-defined. Another challenge is the difference in money available to fund product R&D and money available to fund worker, consumer, and environmental risk research. A final challenge is commercialization of university research projects (spin-offs, private research parks, etc.). Nanotechnology research projects often involve an intensification of this trend. Historically, university research has been regulated differently than commercial research. However, as the boundary between universities and commercial firms blurs, the regulation of university research needs to be rethought.

5. Participation and transparency in standards-setting

Historically, both private and governmental standards-setting bodies have worked with scientific experts and businesses to construct standards. This approach has generated robust national and international standards regimes that have simultaneously advanced and protected proprietary interests while facilitating global commerce and trade. However, this approach is coming under increasing public scrutiny. First, the level and nature of risk that consumers and workers find acceptable may be different from that which businesses and experts consider appropriate. Second, the non-public nature of some standards development can create an impression of collusion and secrecy between industry, experts, and government that can undermine public confidence in standards and standards-setting bodies. Discussion centered on three interrelated themes, including whether and how: (1) public participation should be increased, (2) limits should be placed on such participation, and (3) standards-setting processes can be made more transparent.

Defining and operationalizing the concept of standards
To increase public participation in standards development and implementation, the concept of standards will first have to be defined and operationalized so that the participants are responding to the same basic idea. Key dimensions for clarification include:

- Standards vs. standardization. The terms "standards" and "standardization" refer to distinct concepts. For example, "standards" may be used either to standardize or to differentiate among products, processes, and procedures. Participants in standards-setting processes must be made aware of this distinction so they clearly understand the intended purposes, outcomes, and potential consequences of their participation.

- Formal vs. informal dimension. One dimension that must be clarified immediately is the degree of formality of the standards in question. For example, where along a continuum extending from legally binding restriction and technical proscription at the formal end to social convention at the informal end are such discussions to occur? Traditionally, the more formal standards dimensions provide less room for full public participation; conversely, social conventions are by nature negotiated through open and transparent public interaction.

- Public vs. private dimension. The same may be said of public versus private standards, where private standards are typically negotiated by less public or participatory means.

- Technical vs. strategic dimensions. The very concept of standards needs to be presented publicly as a socially negotiated and strategic phenomenon, rather than solely as the specification of technical attributes or criteria. Standards first need to be recognized as strategic devices that are negotiated among and reflect the interests of participating groups. Standards are thus simultaneously technical and social phenomena that both reflect and are responsive to the broader participation of potentially affected groups.

Defining and identifying potentially affected groups
The bigger questions involve how to identify who the potentially affected groups are, the preferred participatory processes once they have been identified, and clarifying the goals of the process. Some of the key questions that will have to be addressed up front in each of these areas include:

- *Identification* Are these demographic categories of people? Are they defined geographically, socially, culturally, by spheres of economic activity? Do they "self identify," or are they identified by others? And how do companies, private industries, etc., fit into this mix?

- *Process* Once potentially affected populations have been defined and identified, is their participation a function of attending formal standards-setting events, or is it incumbent upon standards-setting organizations to engage in outreach to obtain information from these groups? And in any of these cases, how does one know if one has been successful? The process of participation

must be set to meet the expectations of those who are to participate in that process.

- *Goals* What are the standards for participation in standards-
setting processes? What principles guide the process? What is
the goal? "Better" decisions? Broader representation in decision-
making, regardless of the quality of those decisions? Equitable
distribution of impacts, costs, benefits? Greatest good for the
greatest number of people? Economic efficiency? These things
will need to be agreed upon, or if not agreed upon, then they will
need to form the basis for public discourse. Workshop attendees
identified numerous reasons to pursue public participation in
standards-setting processes, including: (a) it is the right thing to
do, (b) it fosters public trust in standards-setting, (c) it can lead to
greater public protection from unforeseen risks given less participatory processes, and (d) it provides for greater public insight
into regulation. The public may have questions that the regulators
have not considered. Regulation is a long-term affair while public engagement is too often seen as something to be tacked on at
the end of the process. At the same time we need to recognize
that this can slow the regulatory process. One might want to
engage in different forms of participation for different reasons.
There is a need to be sensitive to culturally appropriate forms of
participation. For example, experience conducting public participation among Amish and Native American communities, where
collective decisions are often framed through the counsel of
respected elders, suggests that "one person, one vote" models are
not universally accepted, nor are random samples or statistical
representation necessarily desirable (Stone, 2001).
- *Incentives* From a company perspective there has to be a competitive aspect. There needs to be a clear benefit for encouraging
broad participation. If this cannot be established, then there is little incentive to do it. Then it comes down to being required, or
forced to do so, and this is not always desirable in the absence of
clear benefit. An economic basis for participation must be established. Is it going to be better than what the market would provide? How is this made clear?

Preferred models of participation

Many models exist for public participation and may be adapted in
one way or another to meet the needs of participation in standards-
setting processes as well as the expectations of the participants. The

International Association for Public Participation is a good repository of such information (www.iap.org). Models may range from highly centralized events, such as public hearings, to highly decentralized processes, such as community extension services. These may be highly facilitated/mediated or analytical/deliberative events. In mediated processes, participants resolve a dispute on their own without any "decision" being made by a chairperson or a judge. The resolution in mediation may incorporate agreements on legally irrelevant and often emotionally charged issues. Deliberative processes, on the other hand, are more legalistic or legislative in nature, and typically avoid legally irrelevant and emotionally charged issues. Some examples of successful models used in other contexts include the following:

- Nano Jury's "Mutualistic Engagement" model, builds upon the UK experience with its "GM Nation" effort and is presently being applied to public engagement around emerging nanotechnologies (www.nanojury.org).
- The South Carolina Citizens' School of Nanotechnology is a model of engagement that is particularly well-adapted to public education on nanotechnology (nsts.nano.sc.edu).
- The USDA Cooperative State Research, Education, and Extension Service (CSREES) provides information concerning decentralized extension-based approaches to community outreach and education that are broadly applicable to public dialogues concerning nanotechnology standards (crees.usda.gov).
- The Risk Perception Mapping (RPM) model, developed primarily for social assessment in nuclear waste facility siting, is a decentralized and ethnographic approach that may work well in assessing the potential social impacts associated with siting nanotechnology manufacturing facilities (Stoffle *et al.*, 1991, 1993).

Less clear is how such models would work within standards-setting processes. One thing CODEX has done for international NGOs is to develop standards through electronic workshops rather than physical working groups. Perhaps a hybrid model mixing and matching various elements of these could be developed for specific standards-setting processes.

Public meetings are insufficient

Public meetings provide a venue where people can publicly express themselves, make impassioned pleas on behalf of their communities, demonstrate their commitment and dedication to and concern for their community's well-being, but they do not necessarily provide information regarding the distribution of concerns among a population. One cannot assume a speaker speaks for a community of interest, and the claims made at such meetings should not necessarily be considered public participation.

Drivers of information/insight gained through public participation

What are the drivers of the information being sought through public participation, and what insights are to be gained through such processes?

- *Risk perception* The social impacts literature suggests that such impacts occur to the extent that people perceive themselves to be at risk from something. Risk perception is an important driver in standards participation. For example, one might ask what risks and impacts do potentially affected groups associate with the phenomenon around which standards are being developed, and perhaps more importantly, what are the modes of risk impact.
- *Risk perception analogs* The public perceives risks of new technologies through experience with applications of earlier technologies. Illustrative examples are failures such as Chernobyl, Exxon Valdez, and Bhopal, where the concern was not with the technology per se, but rather with management of the technology. This introduces a new dimension to "risk identification" and management, extending it beyond purely technical considerations and into the realm of social experience with analogous technologies and projects. It also introduces issues surrounding public trust in the institutions charged with managing the risks associated with the technology, or with a specific project or application of it.
- *Trust* Bernard Barber's (1983) work on trust is instructive here. Barber links concepts of trust with public expectations about the future, specifically, (a) "the persistence and fulfillment of the natural and moral social orders," of (b) "technically competent role performance," and that (c) "partners in interaction will carry out their duties in certain situations to place others' interests before their own," what he calls "fiduciary responsibility." Public participation is often marked by a disjuncture among these expectations, particularly (b) and (c), where scientists and technical "experts" typically frame their risk discussions around assumptions and demonstrations of technical competency (e.g. "trust us because we are technically competent"), while potentially affected publics

typically frame the issue in terms that include but extend beyond technical competency to encompass the "fiduciary responsibility" they see as inherent to risk management (e.g. "can this institution be trusted to place broader public interests above its own immediate concerns?"). As such, Barber's work is instructive concerning potential disjunctures in public participation in standards-making. How might these be integrated to build public trust in not only the standard itself, but the process through which it was developed, and indeed perhaps the technology itself?

In this sense, social experience with nuclear power, biotechnology, wireless communications, etc., will be helpful for understanding the kinds of risk perception analogs that will likely drive public participation around nanotechnology, including the development and implementation of standards in this area. This may not generate "better" standards per se as much as a range of "different" standards, niche standards that respond to the values and expectations of distinct communities of interest.

Risk communication

Although education is important to an informed public dialogue on nanotechnology, the participatory issue should not exclusively be about technical understanding but equally about social understanding of how a person and her or his social network stands to be affected. Decision-makers (standards-setters) must remain open to being educated by participants about the social contexts of their concerns—contexts that ultimately will have to be addressed in the standards that are promulgated. Risk communication is a two-way street and must occur early in the process to help frame social contexts of risk perception. This will help clarify the preferred subjects of risk assessments and socially appropriate risk management options for decision-makers. Standards-setting processes have to be collaborative, and participation is a vehicle for that. Otherwise, one has the old linear model of experts deciding what counts and the public reacting to their decisions. The key will be translating the processes used into socially responsive policies and standards—if that is the goal. The International Risk Governance Council is a good example of an organization that deals with such issues, and it has recently released a white paper on nanotechnology risk governance that is informative for public participation in nanotechnology standards development (IRGC, 2007).

Limits of/to participation

There should be procedural but not substantive limits to participation. Some of the key issues to be considered include:

- *Proprietary interests of participants* These must somehow be protected. Multiple layers of proprietary interests are likely to influence the process. Moreover, transparency itself may actually be seen as a disincentive to participate, particularly if transparency means that trade secrets or other proprietary information could be publicly revealed. This is a timing issue: done too early in the process could have the opposite of the desired effect. Then the process is likely to be seen as dishonest rather than transparent.
- *Human subjects* There is a need for full disclosure of purpose and use of information obtained through participation. In addition, there must be assurances that data will be used in only certain ways. Some key questions that must be addressed are: How can this be ensured? Who is responsible for situations where such disclosures are violated? How are disputes arising from this to be adjudicated?
- *Saturation, co-optation, and accountability* There is a need for streamlining among all the many stakeholder organizations and potentially affected groups. Participation can be costly and time-consuming. In the US it tends to be industry representatives who have the financial support to attend these events. If there are 30 meetings a year, which are the ones most worth attending, in terms of both affordability and ability to influence outcomes? Answering this question presumes that one can anticipate outcomes. Yet, one cannot know whether a specific meeting is the one in which it is most worth participating. The participation "market" can quickly become saturated, with no clear direction about to whom to turn or toward what ends the process will lead. For example, the interests being served will likely be reflective of the interests of those who coordinate and implement the event. The perception that one group's interests can be co-opted by another's is a potential disincentive to participation. It creates a cynical perception of the process and ultimately of the decisions reached. Yet, someone has to make these decisions, and not every interest will be equally served. This raises related issues regarding those interests consistently underserved in standards-setting processes.
- *Scale—local to international* Scale presents another potential limit to participation—not everyone can or perhaps even should

participate. Can local input be scaled up to national and international dialogues, and vice versa? This could become quite daunting and the complexity itself could serve as a disincentive to scaling up participation.

- *Equity and social justice* At some point decisions must be made. One has to have milestones for progress in decision-making, and yet at the same time there are many publics that are difficult to reach. Equity issues may be sacrificed for the sake of expediency. Simple "majority rules" solutions may not adequately address equity, particularly in instances where the "minority" views are consistently discounted. There is a need to protect minority perspectives from a "tyranny of the masses," a social justice dimension that should not be overlooked, but often is. In this sense, current limits on participation may actually have to be expanded in order to allow for greater discussion of equity and social justice. But determining who will make that call, and on what grounds, is the subject of broad public dialogue. Key questions in this area include: (a) how equity issues can be addressed procedurally, and (b) whether participants are willing to accept outcomes or decisions that they might consider inappropriate. This condition will have to be understood by all parties going into the process. Otherwise the process will unravel and simply become the domain of a self-selected subset of interests. This cannot legitimately be called "participation."

Toward greater transparency in standards-setting processes
In standards-setting, no open-source mechanisms exist in which the public can clearly see the process and is welcome to be part of it. This would assume an educated public, at least to some degree, so a truly open-source mechanism could only be implemented upon the shoulders of a broader public education mechanism. However, as noted previously, public education is a two-way communications issue in which the point is not just to educate the public on the technical aspects of standards issues, but to be educated by them concerning the social contexts of their concerns.

- *Formative evaluation* Formative evaluation can help to increase transparency in standards-setting processes. Key evaluation questions concern: (a) establishing a clear definition or understanding of "transparency," specifically its goals and procedural objectives, and (b) how one knows when these have been met. The answers to these questions will likely vary by stakeholder interest, so this dialogue needs to occur independently of the standards-setting process. Answers to these questions will help to establish appropriate models of participation that can then be pursued to maximize transparency as so defined. This should be an iterative or formative process in which outcomes inform the implementation of future processes, that is, in terms of the definitions, goals, and procedural objectives of both transparency and the participatory procedures used to obtain it.

- *Educating the public on nanotechnology* The South Carolina Citizens' School of Nanotechnology provides a good example of educating the public about nanotechnology generally, but turnout has been quite low to date. Also, it is unclear how representative of broader publics such techniques actually are, or whether they have even been conceived to address representational issues. Clear representation in educational programs is important—determining who speaks for whom, and how (or whether) each voice in the process can be weighted. The South Carolina program team is presently revising the Citizens' School process to explicitly address these issues.

- *Educating decision-makers on social context* National representatives to Codex and other groups have a responsibility to collect information from their respective publics. But there are no consistent procedures whereby national organizations are expected to interact with them. The information is inconsistent, and there is little guidance regarding how that information will actually be used to help shape standards decisions. A recent report of a joint workshop of the Food and Agriculture Organization (FAO) and the World Health Organization (WHO) indicated that in such cases there must be room for minority opinions and decision-makers must be accountable for why such opinions may not be factored into the decisions reached (FAO/WHO, 2004). In the UK, Defra (Department of Environment, Food, and Rural Affairs) has utilized focus group models to increase public accessibility to nanotechnology discussions, and this could be adapted to increase transparency of nanotechnology standards deliberations, or at least public understanding of how these processes work (www.defra.gov.uk). A public advisory board could work as intermediary between the formal standards-setting process and the multitude of voices that could potentially demand a formal role in the process. The interaction between the advisory board and the public would be very transparent. The board would then take that information to the standards committee. Someone still has to make decisions, though, and the public is left largely to react rather than collaborate on the decisions reached.

Conclusion

Timing

Given the concerns noted above, standard-setting should start early and be seen as a strategic and interactive process. Indeed, all regulation and risk assessment has operated on the basis of incomplete knowledge. Regulators should note that they are doing their best within the current framework, rather than employing a "trust us" approach. This will involve an admission by regulators of how much they do not know. At the same time, it will require a significant increase in funding to identify risks and to engage in formal risk assessments. The NNI reports that 4% (roughly $40 million) of its FY 2006 budget was dedicated to R&D aimed primarily at understanding and addressing the potential risks posed by nanotechnology to health and the environment (NNCO, 2005; PCAST, 2005). According to PCAST (2005), this amount does not include research of a different primary focus but that nonetheless extends knowledge of health and environmental effects of nanomaterials. A recent report of the Woodrow Wilson Center's Project on Emerging Nanotechnologies, however, contests the NNI's $40 million figure, suggesting that only $11 million is actually geared to "risk relevant" research. Still, semantic issues aside, Wilson Center researchers claim neither amount is adequate and instead recommend a minimum funding increase of $50 million over each of the next 2 years (Maynard, 2006a, 2006b). If risk assessment moves slowly while product development moves rapidly, we may be more likely to experience a calamitous incident. Initially, standards should focus on laboratory research and research institutions, as well as on reporting of incidents. Simultaneously, funding for risk research should be increased. Since rule-making is a lengthy process, specific deadlines should be established for preliminary standards. It should begin with conservative and inclusive standards, so as to reduce exposure to potential hazards, but should not be so conservative as to bring research to a halt. The later standards emerge, the more vulnerable everyone involved will become. Timing is also related to transparency in that a rapid decision about the need for process or product standards may influence transparency and disclosure requirements. This will in turn affect public opinion about nanotechnology and the level of public control over decisions.

Product vs. process

Standard-setting for nanotechnologies is going on in many nations. While the creation of globally acceptable international standards for nanotechnology products, processes, research, environment, and health and safety is still far off, cooperation and information exchange need not wait until individual national standards are established. Nor should national standards be avoided because one is worried about competition, or because international standards are insufficiently developed. Indeed, for global trade at the very least everyone must use the same language. Such global collaboration can include identification of risk factors, means for public engagement, public opinion polling, and public education. In addition, trade involving certain nano-products (e.g. radio frequency identification devices) will need to take into account and develop appropriate methods for preserving privacy with respect to the vast amounts of data generated in the process, as well as various obligations under the World Trade Organization.

Product/process standards may also be affected by the initial ordering of the standards (i.e. the sequence in which product and process standards are introduced). There was some division among the participants, with some arguing that processes should be subject to standards first and others arguing that product standards should be developed first.

It was reiterated that the goals for and safety of the technologies will affect the standards. In this way, the standards will depend on the nanotechnology in question. There was disagreement as to how cautious to be. Some argued that technologies should be treated as potentially harmful until demonstrated to be safe, while others argued that we must experiment both with the technologies and with the standards. In the end, the practical products and processes that are in use or about to be used will force us to establish standards.

Process and product standards should vary by sector only if what works within that sector differs from other sectors. In other words, if there are similar goals across sectors, then there should be consistency. Imposition of a one size fits all approach is unlikely to succeed. The ability to experiment should be preserved.

International harmonization

Every debate over harmonization is about national differences in beliefs about workers, the environment, competitiveness, etc. At the same time, public demand for transparency in the harmonization of standards is likely to be a driving force. It might also be useful to talk about what values we wish to promote. Coordination with the United Nations Development Programme's Millennium Development Goals (www.undp.org), for example, might be worth pursuing.

Integration of operational standards

One must ask whether standards for research and development should be integrated with consumer, worker, and environmental standards or not. Such a move may slow R&D, but with the unique concerns associated with nanotechnology there is reason to consider this option. If nations or international bodies move in this direction, it will be necessary to identify the forces that prevent effective integration and work to overcome them. Furthermore, overcoming existing power, hierarchy, and domination in the standards/regulation development process is important to attain publicly acceptable results. One way to address this is by developing ways for NGOs and other citizen groups to play a meaningful role.

Participation/transparency

Public education and engagement should proceed concurrently with the standards-setting process. Currently, public awareness of nanotechnologies is limited, but growing. More public education is needed if a true dialogue is to take place. The more dialogues that take place, the more awareness will increase. In 2003, Congress mandated education on this issue (United States Congress, 2003). This bill authorizes appropriations for nanoscience, nanoengineering, and nanotechnology research, and for other purposes—but little has happened to date. However, successful models of rapid public education do exist, including that of the Cooperative Extension Service. These should be used as needed. Public interest organizations should be constructively involved in the process from its inception, although limits of time and resources make it impossible to include "all interested parties." Encouraging high levels of participation will likely reduce both negative and adversarial aspects of the process. Potentially affected parties need to be identified rapidly, so that means for mitigation can be identified more easily.

Sources for standards and nanotechnology

Following is a list of key links and references identified by participants during the workshop or otherwise cited in this document.

- How EPA statutes may be used for regulation of nanotechnology: http://www.abanet.org/environ/nanotech/
- The UK's Department of Environment, Food, and Rural Affairs (Defra) has utilized focus group models to increase public accessibility to nanotechnology discussions. This could be adapted to increase transparency of nanotechnology standards deliberations: http://www.defra.gov.uk/
- The Brazilian Agricultural Research Corporation (EMBRAPA) seeks feasible solutions for the sustainable development of Brazilian agribusiness through knowledge and technology generation and transfer: Further information (in English) regarding EMBRAPA programs and publications may be found at: http://www.embrapa.br/English/index_html/mostra_documento
- Former EPA administrator J. Clarence Davies explains why he believes nanotechnology-specific legislation is necessary: http://www.eande.tv/transcripts/?date=030106
- In 2001, the EPA convened a National Dialogue on Public Involvement in EPA Decisions, which addresses participation and transparency issues from an EPA perspective. Documents pertaining to this process may be found at: http://www.networkdemocracy.org/epa-pip/
- Based largely on input received through its National Dialogue on Public Involvement in EPA Decisions, the EPA released its Public Involvement Policy in 2003, the details of which may be found at: http://www.epa.gov/publicinvolvement/policy2003/

- Institute for Agriculture and Trade Policy (IATP) comment to the FDA on regulated products containing nanotechnology materials: http://www.environmentalobservatory.org/library.cfm? refid=89139
- The International Association for Public Participation (IAP2) is a good repository of information on models of public participation: http:// www.iap2.org/index.cfm
- An examination of health and environmental safety issues is on the webpage of the International Risk Governance Council: http://www.irgc.org/irgc/projects/nanotechnology/
- The International Risk Governance Council's White Paper on Nanotechnology Risk Governance (June, 2006): http://www. irgc.org/irgc/_b/contentFiles/IRGC_white_paper_2_PDF_final_ version.pdf
- Nano Jury's "Mutualistic Engagement" model builds upon the UK experience with its "GM Nation" effort and is presently being applied to public engagement around emerging nanotechnologies (see, e.g., http://www.nanojury.org/intro_mutual.htm)
- The South Carolina Citizens' School of Nanotechnology is a model of engagement that is particularly well-adapted to public education on nanotechnology: http://nsts.nano.sc.edu/outreach. html
- USDA Cooperative State Research, Education, and Extension Service (CSREES) outlines decentralized extension-based approaches to community outreach and education that are broadly applicable to public dialogues concerning nanotechnology standards: http://www.csrees.usda.gov/qlinks/extension.html
- The United Nations Development Programme's "Millennium Development Goals" provides a working example for international harmonization around global interests of shared concern: http://www.undp.org/mdg/

Endnotes

1. The IFAS mission is to raise questions about such fundamental issues as equity, fairness and transparency of food and agricultural standards at the local, national and international levels.
2. See Appendix III for a list of acronyms and associated definitions as used in this document.

References

ASTM Committee E-56 (2006). *E2456-06, Standard Terminology Relating to Nanotechnology*. ASTM International.

Barber, B. (1983). *The Logic and Limits of Trust*. Rutgers University Press.

FAO/WHO, Food and Agriculture Organization of the United Nations and the World Health Organization (2004). Joint FAO/WHO Workshop on the Provision of Scientific Advice to Codex and Member Countries. 27–29 January 2004. FAO/WHO.

Mattoso, L. H. C., Medeiros, E. S., and Martin-Neto, L. (2005). A revolução nanotecnológica e o potencial para o agronegócio. (The nanotechnology revolution and the potential application in agribusiness). *Revista de Política Agrícola* **14**, 38–48 (in Portuguese).

Maynard, A. D. (2006a). *Nanotechnology: A Research Strategy for Addressing Risk*. Woodrow Wilson International Center for Scholars, Project on Emerging Nanotechnologies.

Maynard, A. D. (2006b). Safe handling of nanotechnology. *Nature* **444**, 267–269 (Commentary, November 16, 2006).

NNCO, National Nanotechnology Coordination Office (2005). The National Nanotechnology Initiative: Research and Development Leading to a Revolution in Technology and Industry. Supplement to the President's 2006 Budget.

OSTP, Office of Science and Technology Policy (1986). National Nanotechnology Coordination Office. Coordinated Framework for Regulation of Biotechnology; Announcement of Policy and Notice for Public Comment. *Federal Register* **51**(123), 23302–23350.

PCAST, President's Council of Advisors on Science and Technology (2005). *The National Nanotechnology Initiative at Five Years: Assessment and Recommendations of the National Nanotechnology Advisory Panel*. Office of Science and Technology Policy.

Rattner, H. (2005). Nanotecnologia e a política de ciência e tecnologia (Nanotechnology and the science and technology policy). *Passages de Paris* **2**, 180–188 (in Portuguese).

Roco, M. C. (2003). *Government Nanotechnology Funding: An International Outlook*. Nanoscale Science, Engineering, and Technology Subcommittee. National Science Foundation.

Stoffle, R. W., Traugott, M. W., Stone, J. V., McIntyre, P. D., Jensen, F. V., and Davidson, C. D. (1991). Risk perception mapping: using ethnography to define the locally affected population for a low-level radioactive waste storage facility in Michigan. *Am Anthropol* **93**, 611–635.

Stoffle, R. W., Stone, J. V., and Heeringa, S. (1993). Mapping risk perception shadows: defining the locally affected population for a low-level radioactive waste facility in Michigan. *Environ Profess* **15**, 316–333.

Stone, J. V. (2001). Risk perception mapping and the Fermi II nuclear power plant: toward an ethnography of social access to public participation in great lakes environmental management. *J Environ Sci Policy* (special issue on *Environmental Knowledge, Rights, and Ethics: Co-managing with Communities*) **4**, 205–217.

Stone, J. V. and Wolfe, A. K. (2006). Introduction to "Nanotechnology in society: atlas in wonderland?" *Practicing Anthropology* (special issue on *Nanotechnology in Society*) **28**, 2–5.

United States Congress (2003). Public Law 108–153: A bill to authorize appropriations for nanoscience, nanoengineering, and nanotechnology research, and for other purposes. USC 12/3/2003.

Internet references

IFAS, Institute for Food and Agricultural Standards, Michigan State University, 2007. An Issues Landscape for Nanotechnology Standards: Report of a Workshop, http://ifas.msu.edu/NS WorkshopReport.pdf

IRGC, International Risk Governance Council (2007). *Nanotechnology Risk Governance.* http://www.irgc.org/irgc/IMG/ projects/ IRGC_white_paper_2_PDF_final_version.pdf

Appendix III
List of
Abbreviations

ANSI	American National Standards Institute
ASME	American Society of Mechanical Engineers
ASTM	American Society for Testing and Materials
CODEX	Codex Alimentarius Commission
CSREES	Cooperative State Research, Education and Extension Service
Defra	UK Department of Environment, Food and Rural Affairs
EPA	US Environmental Protection Agency
FAO	Food and Agriculture Organization
FDA	US Food and Drug Administration
GMOs	Genetically modified organisms
IFAS	Institute for Food and Agricultural Standards
ISO	International Organization for Standardization
MSU	Michigan State University
NABC	National Agricultural Biotechnology Council
NAS	National Academy of Sciences
NGO	Non-governmental organization
NNCO	National Nanotechnology Coordination Office
NNI	National Nanotechnology Initiative
NSF	National Science Foundation
OECD	Organization for Economic Co-operation and Development
OSHA	Occupational Safety and Health Administration
OSTP	Office of Science and Technology Policy
PCAST	President's Council of Advisors on Science and Technology

What Can Nanotechnology Learn from Biotechnology?
ISBN: 978-012-373990-2

PSRAST	Physicians and Scientists for Responsible Science and Technology
R&D	Research and development
RAFI	Rural Advancement Foundation International
RFID	Radio-frequency identification device
RPM	Risk perception mapping
USDA	United States Department of Agriculture
WHO	World Health Organization

Appendix IV Participants at First International IFAS Conference on Nanotechnology "What Can Nano Learn from Bio?"

Held at Michigan State University, Kellogg Hotel and Conference Center, East Lansing Michigan, October 26–27, 2005

Jim Aidala	Bergeson and Campbell
Fritz Allhoff	Western Michigan University
John Bedz	Michigan Small Tech Association
David Berube	University of South Carolina
Lori Bestervelt	NSF International
Les Bourquin	Michigan State University
Mason Bradbury	Michigan State University*
Jeffrey Burkhardt	University of Florida

What Can Nanotechnology Learn from Biotechnology?
ISBN: 978-012-373990-2

Larry Busch	Michigan State University
Kenneth David	Michigan State University
Brady Deaton	University of Guelph
Thomas Deits	Lansing Community College
Brian Depew	Michigan State University**
Thomas Dietz	Michigan State University
Abby Dilley	RESOLVE
George Gaskell	London School of Economics
Hans Geerlings	Shell Research and Technology Centre
Mickey Gjerris	Danish Centre for Bioethics
William Hannah	Michigan State University**
Jaydee Hanson	International Center for Technology Assessment
Mrill Ingram	La Follette School of Public Affairs
Michael Knox	LLC
Jennifer Kuzma	University of Minnesota
Sue LaVigne	Gerber Products Co
Richard Lempert	National Science Foundation
Jacqueline Leshkevich	LSB-Research Services Division
John (Jack) Lloyd	Michigan State University
Phil Macnaghten	Lancaster University
Alan McHughen	University of California-Riverside
Zahra Meghani	Michigan State University**
Manish Mehta	National Center for Manufacturing Sciences
Margaret Mellon	Union of Concerned Scientists
Sonia Miller	Converging Technologies Bar Association
Kirsty Mills	Center For High Technology Materials
Sho Ngai	Michigan State University**
Norbismi Nordin	Michigan State University**
Chris Phoenix	Center for Responsible Nanotechnology
Susanna Priest	University of South Carolina
Erin Pullen	Michigan State University*
Priscilla Regan	National Science Foundation
Jason Robert	ASU Center for Biology and Society
Sue Selke	Michigan State University
Steven Slack	The Ohio State University/OARDC
David Sparling	University of Guelph
Chris Steele	NSF International (The Toxicology Group)

John Stone	Michigan State University
Meghan Sullivan	Michigan State University**
Keiko Tanaka	Michigan State University*
Keiko Tanaka	University of Kentucky
Deepa Thiagarajan	Michigan State University
Paul Thompson	Michigan State University
Peter VerHage	University of Minnesota
Amy Wolfe	Oak Ridge National Lab

---

*Undergraduate Student Assistant
**Graduate Student Assistant

Appendix V
Participants in the "Standards for Nanotechnology" Workshop, 2006

Fritz Allhoff	Western Michigan University, Philosophy
Evangelyn Alocilja	MSU—Biosystems and Agriculture Engineering
Les Bourquin	MSU—Food Science & Human Nutrition
Larry Busch	MSU—IFAS
Flora Chow	US Environmental Protection Agency
Renata Clarke	Food & Agriculture Organization of the United Nations (FAO)
Robert Clarke	MSU—Packaging
Kenneth David	MSU—Anthropology
Odílio Benedito Garrido de Assis	CNPDIA/EMBRAPA Agricultural Instrumentation
Brady Deaton, Jr.	University of Guelph, Agricultural Economics
Thomas Deits	Lansing Community College Science Dept.
Brian Depew	MSU—Sociology

What Can Nanotechnology Learn from Biotechnology?
ISBN: 978-012-373990-2

Mark Duvall	Dow Chemical Company
William Hannah	MSU—Philosophy
Jaydee Hanson	International Center for Technology Assessment
Barbara Herr Harthorn	Center for Nanotechnology in Society: University of California-Santa Barbara
Steve Holcombe	Pardalis Inc.
John Koehr	American Society of Mechanical Engineers (ASME)
Kristen Kulinowski	Center for Biological and Environmental Nanotechnology, Rice University
Jennifer Kuzma	Center for Science, Technology & Public Policy
John (Jack) Lloyd	MSU—Mechanical Engineering
Michele Mekel	Converging Technologies Bar Association (CTBA)
Doug Meyer	International Union of Food Workers
Kirsty Mills	Center for High Technology Materials at University of New Mexico
Julia Moore	Woodrow Wilson International Center for Scholars
Stephen Mooser	Retail, Wholesale and Department Store Union
Ed Munn	University of South Carolina
Mark Nelson	Grocery Manufacturers of America (GMA)
Sho (Lisa) Ngai	MSU—Mechanical Engineering
Norbismi Nordin	MSU—Packaging
Steven Pueppke	MSU—Michigan Agriculture Experiment Station
Susan Selke	MSU—Packaging
John V. Stone	MSU—IFAS
Steve Suppan	Institute for Agriculture and Trade Policy (IATP)
Toby Ten Eyck	MSU—National Food Safety & Toxicology Center
Deepa Thiagarajan	MSU—Institute for International Agriculture
Jim Thomas	ETC Group

Paul Thompson	MSU—W.K. Kellogg Chair in Agricultural, Food and Community Ethics
Jameson Wetmore	The Robert M. La Follette School of Public Affairs
Jane Wilson	National Sanitation Foundation (NSF) International

Index

Food science and technology

International Series

Maynard A. Amerine, Rose Marie Pangborn, and Edward B. Roessler, *Principles of Sensory Evaluation of Food*. 1965.

Martin Glicksman, *Gum Technology in the Food Industry*. 1970.

Maynard A. Joslyn, *Methods in Food Analysis*, second edition. 1970.

C. R. Stumbo, *Thermobacteriology in Food Processing*, second edition. 1973.

Aaron M. Altschul (ed.), *New Protein Foods:* Volume 1, *Technology, Part A*—1974. Volume 2, *Technology, Part B*—1976. Volume 3, *Animal Protein Supplies, Part A*—1978. Volume 4, *Animal Protein Supplies, Part B*—1981. Volume 5, *Seed Storage Proteins*—1985.

S. A. Goldblith, L. Rey, and W. W. Rothmayr, *Freeze Drying and Advanced Food Technology*. 1975.

R. B. Duckworth (ed.), *Water Relations of Food*. 1975.

John A. Troller and J. H. B. Christian, *Water Activity and Food*. 1978.

A. E. Bender, *Food Processing and Nutrition*. 1978.

D. R. Osborne and P. Voogt, *The Analysis of Nutrients in in Foods*. 1978.

Marcel Loncin and R. L. Merson, *Food Engineering: Principles and Selected Applications*. 1979.

J. G. Vaughan (ed.), *Food Microscopy*. 1979.

J. R. A. Pollock (ed.), *Brewing Science*, Volume 1—1979. Volume 2—1980. Volume 3—1987.

J. Christopher Bauernfeind (ed.), *Carotenoids as Colorants and Vitamin A Precursors: Technological and Nutritional Applications*. 1981.

Pericles Markakis (ed.), *Anthocyanins as Food Colors*. 1982.

George F. Stewart and Maynard A. Amerine (eds.), *Introduction to Food Science and Technology*, second edition. 1982.

Malcolm C. Bourne, *Food Texture and Viscosity: Concept and Measurement*. 1982.

Hector A. Iglesias and Jorge Chirife, *Handbook of Food Isotherms: Water Sorption Parameters for Food and Food Components*. 1982.

Colin Dennis (ed.), *Post-Harvest Pathology of Fruits and Vegetables*. 1983.

P. J. Barnes (ed.), *Lipids in Cereal Technology*. 1983.

David Pimentel and Carl W. Hall (eds.), *Food and Energy Resources*. 1984.

Joe M. Regenstein and Carrie E. Regenstein, *Food Protein Chemistry: An Introduction for Food Scientists*. 1984.

Maximo C. Gacula, Jr., and Jagbir Singh, *Statistical Methods in Food and Consumer Research*. 1984.

Fergus M. Clydesdale and Kathryn L. Wiemer (eds.), *Iron Fortification of Foods*. 1985.

Robert V. Decareau, *Microwaves in the Food Processing Industry*. 1985.

S. M. Herschdoerfer (ed.), *Quality Control in the Food Industry*, second edition. Volume 1—1985. Volume 2—1985. Volume 3—1986. Volume 4—1987.

F. E. Cunningham and N. A. Cox (eds.), *Microbiology of Poultry Meat Products*. 1987.

Walter M. Urbain, *Food Irradiation*. 1986.

Peter J. Bechtel, *Muscle as Food*. 1986.

H. W. -S. Chan, *Autoxidation of Unsaturated Lipids*. 1986.

Chester O. McCorkle, Jr., *Economics of Food Processing in the United States*. 1987.

Jethro Japtiani, Harvey T. Chan, Jr., and William S. Sakai, *Tropical Fruit Processing*. 1987.

J. Solms, D. A. Booth, R. M. Dangborn, and O. Raunhardt, *Food Acceptance and Nutrition*. 1987.

R. Macrae, *HPLC in Food Analysis*, second edition. 1988.

A. M. Pearson and R. B. Young, *Muscle and Meat Biochemistry*. 1989.

Marjorie P. Penfield and Ada Marie Campbell, *Experimental Food Science*, third edition. 1990.

Leroy C. Blankenship, *Colonization Control of Human Bacterial Enteropathogens in Poultry*. 1991.

Yeshajahu Pomeranz, *Functional Properties of Food Components*, second edition. 1991.

Reginald H. Walter, The Chemistry and Technology of Pectin. 1991.

Herbert Stone and Joel L. Sidel, Sensory Evaluation Practices, second edition. 1993.

Robert L. Shewfelt and Stanley E. Prussia, *Postharvest Handling: A Systems Approach*. 1993.

R. Paul Singh and Dennis R. Heldman, *Introduction to Food Engineering*, second edition. 1993.

Tilak Nagodawithana and Gerald Reed, *Enzymes in Food Processing*, third edition. 1993.

Dallas G. Hoover and Larry R. Steenson, *Bacteriocins*. 1993.

Takayaki Shibamoto and Leonard Bjeldanes, *Introduction to Food Toxicology*. 1993.

John A. Troller, *Sanitation in Food Processing*, second edition. 1993.

Ronald S. Jackson, *Wine Science: Principles and Applications*. 1994.

Harold D. Hafs and Robert G. Zimbelman, *Low-fat Meats*. 1994.

Lance G. Phillips, Dana M. Whitehead, and John Kinsella, *Structure-Function Properties of Food Proteins*. 1994.

Robert G. Jensen, *Handbook of Milk Composition*. 1995.

Yrjö H. Roos, *Phase Transitions in Foods*. 1995.

Reginald H. Walter, *Polysaccharide Dispersions*. 1997.

Gustavo V. Barbosa-Cánovas, M. Marcela Góngora-Nieto, Usha R. Pothakamury, and Barry G. Swanson, *Preservation of Foods with Pulsed Electric Fields*. 1999.

Ronald S. Jackson, *Wine Science: Principles, Practice, Perception*, second edition. 2000.

R. Paul Singh and Dennis R. Heldman, *Introduction to Food Engineering*, third edition. 2001.

Ronald S. Jackson, *Wine Tasting: A Professional Handbook*. 2002.

Malcolm C. Bourne, *Food Texture and Viscosity: Concept and Measurement*, second edition. 2002.

Benjamin Caballero and Barry M. Popkin (eds), *The Nutrition Transition: Diet and Disease in the Developing World*. 2002.

Dean O. Cliver and Hans P. Riemann (eds), *Foodborne Diseases, second edition*. 2002.

Martin Kohlmeier, *Nutrient Metabolism*, 2003.

Herbert Stone and Joel L. Sidel, *Sensory Evaluation Practices*, third edition. 2004.

Jung H. Han, *Innovations in Food Packaging*. 2005.

Da-Wen Sun, *Emerging Technologies for Food Processing*. 2005.

Hans Riemann and Dean Cliver (eds) *Foodborne Infections and Intoxications*, third edition. 2006.

Ioannis S. Arvanitoyannis, *Waste Management for the Food Industries*. 2008.

Ronald S. Jackson, *Wine Science: Principles and Applications*, third edition. 2008.

Da-Wen Sun, *Computer Vision Technology for Food Quality Evaluation*. 2008.

Kenneth David, *What Can Nanotechnology Learn From Biotechnology?* 2008.

Elke Arendt, *Gluten-Free Cereal Products and Beverages*. 2008.